Einstein's Relativity

AF307199

Fred I. Cooperstock · Steven Tieu

Einstein's Relativity

The Ultimate Key to the Cosmos

 Springer

Dr. Fred I. Cooperstock
University of Victoria
Victoria, BC
Canada

Dr. Steven Tieu
Systemware Innovation
Toronto, ON
Canada

ISBN 978-3-642-44617-7   ISBN 978-3-642-30385-2 (eBook)
DOI 10.1007/978-3-642-30385-2
Springer Heidelberg New York Dordrecht London

© Springer-Verlag Berlin Heidelberg 2012
Softcover reprint of the hardcover 1st edition 2012
This work is subject to copyright. All rights are reserved by the Publisher, whether the whole or part of the material is concerned, specifically the rights of translation, reprinting, reuse of illustrations, recitation, broadcasting, reproduction on microfilms or in any other physical way, and transmission or information storage and retrieval, electronic adaptation, computer software, or by similar or dissimilar methodology now known or hereafter developed. Exempted from this legal reservation are brief excerpts in connection with reviews or scholarly analysis or material supplied specifically for the purpose of being entered and executed on a computer system, for exclusive use by the purchaser of the work. Duplication of this publication or parts thereof is permitted only under the provisions of the Copyright Law of the Publisher's location, in its current version, and permission for use must always be obtained from Springer. Permissions for use may be obtained through RightsLink at the Copyright Clearance Center. Violations are liable to prosecution under the respective Copyright Law.
The use of general descriptive names, registered names, trademarks, service marks, etc. in this publication does not imply, even in the absence of a specific statement, that such names are exempt from the relevant protective laws and regulations and therefore free for general use.
While the advice and information in this book are believed to be true and accurate at the date of publication, neither the authors nor the editors nor the publisher can accept any legal responsibility for any errors or omissions that may be made. The publisher makes no warranty, express or implied, with respect to the material contained herein.

Cover illustration by Travis Morgan, morgantj@gmail.com

Printed on acid-free paper

Springer is part of Springer Science+Business Media (www.springer.com)

All physical theories, their mathematical expressions notwithstanding, ought to lend themselves to so simple a description that even a child could understand them

Attributed to Albert Einstein by Louis de Broglie in Nouvelles perspectives en microphysique (trans. New York; Basic Books, 1962), 184

In memory of Matt, Dani, and Jerry and
In honour of N. V. Tieu and V. Kha

Acknowledgments

We are grateful to our senior editor, Angela Lahee for her considerate guidance and advice throughout the course of our progress. An anonymous reader affiliated with the publisher carefully scrutinized an early draft of the manuscript and provided us with several useful points and corrections. Editorial assistant Claudia Neumann kindly translated the text of the reader into English. We thank them both for their kind efforts. As always, Peter Gary was a constant source of inspiration.

Ruth Cooperstock carefully examined this book for grammatical points and provided us with valuable guidance as to explanations that would appeal to the lay-reader. To the extent that we were able to do so, we followed her sage advice. Her encouragement and devoted support were instrumental in seeing this work through to its completion.

Contents

Chapter 1
Introduction

If you were to stop the first three people that you meet on the street and ask them what they know of Einstein's Relativity, you would likely get answers of the following form:

"I don't know anything about physics. But Einstein was this world-famous genius with a big mop of hair, who wandered around in his slippers and crumpled sweatshirt."

The second might get more technical and say: "I know: $E = mc^2$ and this made the atomic bomb."

The third might reply with

"Oh, there is this rhyme:

There once was a girl named Bright/ Whose speed was much faster than light./ She set out one day/ In a relative way/ And returned on the previous night. It sounds crazy but what do I know?"

Later, we will delve into the real story of the remarkable Ms Bright and her equally remarkable sisters.

Since you are reading this book, you wish to know more about Relativity and well you should. *The essential aim of this book is to bring you, the reader, to a level of appreciation of Einstein's Relativity that you had never previously believed was possible without a great deal of additional training.* This goal is in line with Einstein's exhortation above, but at a necessarily higher level since our aim is more ambitious.

Einstein's Relativity has changed the face of physics and it is unfortunate that so little is known about Relativity beyond the professional scientific community. His theories of Relativity (yes, plural, there are two, there is the general theory that blends into the special theory under the right conditions as we will discuss later) embody science at its very best. They are bold, elegant, imaginative, logically structured and encompass so much of what we do in physics. Above all for science, the Special Theory of Relativity is so well tested as to be accepted almost universally and the general theory, while far more limited in experimental verification, is almost universally regarded as our very best theory of gravity. Part of our goal is to show you

F. I. Cooperstock and S. Tieu, *Einstein's Relativity*,
DOI: 10.1007/978-3-642-30385-2_1, © Springer-Verlag Berlin Heidelberg 2012

that Einstein's General Relativity has an even greater degree of applicability than is generally believed by many.

Most people have an intuitive notion of what physics is but cannot define it precisely. By definition, physics is the scientific study of the properties and interactions of matter and energy. Given its extraordinary scope, physicists are often inclined to wonder how anyone would not wish to study physics. We will take you through a journey that will not even require any physics background to appreciate a goodly part of this book, possibly even more. At the same time, if you still remember or are willing to dust off the simple algebra that you had learned in junior high-school, you will gain a still better appreciation of this gem of science. Insofar as possible, we will use illustrations to bring out the essential ideas. For Special Relativity, diagrams will make otherwise bewildering concepts crystal-clear. As well, such diagrams will be used in a couple of suggestive forms for the curved space of General Relativity. This will reveal in a simple picture, the big bang and the potential for a big crunch of the universe. We will have succeeded in our task if your journey is pleasurable and you come away from it feeling that you have a real appreciation of Einstein's stellar achievements.

We will begin with that other paramount genius in the entire history of physics, Isaac Newton. We will discuss his laws of motion which form the foundation of classical mechanics, laws which served us entirely for our primary endeavours up until the twentieth century and continue to do so to this day.

As a lead-in to Special Relativity, we will take a brief excursion through the basics of electromagnetism since electromagnetic radiation plays such a fundamental role in Relativity as well as in our basic existence. Our approach to Special Relativity will be that of H. Bondi who, under the influence of J.L. Synge, described the essentials of Special Relativity in terms of "spacetime" diagrams. These diagrams are analogous to the blueprints that are used in the construction of a building. Just as blueprints map out pictorially the locations of rooms and fixtures in the space of a building, the spacetime diagrams of Bondi map out the locations and the motions of observers and light rays in spacetime. We will be referring to spacetime frequently as we proceed. Spacetime is that essential new structure with which we will become familiar in Relativity. In Relativity, space and time become unified in a very elegant manner. However, it is not as if, in Relativity, that time is placed on the *same* footing as space, a mistaken notion that many have come to believe. Space and time remain as different concepts but they link in a new mathematical structure, a marriage if you will, that we call "spacetime", and with this union, we will see that time no longer carries that absolute character that it had maintained in physics over the earlier centuries.

Bondi's approach serves as an excellent introduction towards a full appreciation of the nature of spacetime. With his approach, it becomes clear how observers in relative motion perceive each other's length and time intervals differently. As well, the compounding of velocities and the transformation between the space and time coordinates of events as perceived by observers in relative motion are easily derived. Of particular fascination is the so-called "twin paradox", the seeming contradiction that arises when one of two twins takes off on a lengthy journey at very high velocity and returns to find the stay-at-home twin having aged more than the traveler-twin.

One is inclined to ask why it could not be viewed just as well as the stay-at-home twin having made the journey and hence the ageing effect reversed, hence the paradox. This, as well as a missing ingredient in the Bondi approach, will be elucidated for you, the reader.

Since paradoxes in Relativity have been a source of both wonderment and confusion for readers, we have paid considerable attention to them in this book. In addition to the twin paradox, we examine two other paradoxes, each useful in its own way to bring out aspects of Relativity. First, we consider a stick that slides along the ice at high velocity and gets smacked when it lies over a hole in the ice. The stick is viewed as shortened relative to observers at rest on the ice. Thus these observers are seen as readily capable of getting the stick to become submerged under the ice. However the observers who are riding on the stick see the hole in the ice as contracted and hence the hole seemingly too short to enable the stick to fit. Second, we consider a rod that is being carried at high velocity through the door of a barn and being trapped inside the barn with the door closed when the rod is entirely inside. However, relative to the rest frame of the rod, the barn appears shortened and hence the paradox arises as to how the rod could have become trapped in a barn that is too short to accommodate it. For both of these cases, we extract the reality of what actually occurs and how the phenomena appear for each set of observers. Einstein's Relativity does not lead us to logical contradictions.

We will explore the interesting nature of energy and momentum in Relativity, how there emerges a base level of energy for any given amount of matter that exists and how enormous that energy content really is. It is the matter of the famous equation $E = mc^2$. With this equation, we appreciate the actual unification of the concepts of mass and energy in Relativity.

At that point, we introduce the crowning achievement of Einstein's vision, his theory of General Relativity, the incorporation of gravity into Relativity. Here we emphasize that true gravity, produced by matter and fields such as electromagnetism, is not a field as we have come to know other fields in the past but rather it is a manifestation of the curvature of spacetime, a property of a different kind of geometry. This is in line with J. L. Synge who took issue with the Equivalence Principle, the notion that gravity is physically equivalent to an accelerated reference frame. In actuality, it is only an approximate equivalence of effect and not a physical equivalence. This is a frequent source of confusion and we take pains to clarify this important issue. We then go on to develop the equation of motion of a free body in General Relativity, the so-called geodesic equation. In this, we see the essential geometric character of the theory of General Relativity, the linking of physical motion to the paths of extremal distance in curved spacetime. We briefly develop the field equations for General Relativity, the equations to which we are guided by the fundamental physical laws of the conservation of energy and momentum. Many have come to view these Relativity equations as the most fundamental and the most beautiful in all of physics.

At this point, we turn to what is arguably the most important solution of the Einstein equations, the Schwarzschild solution, which describes the gravity produced by any spherically symmetric body. The solution is beautiful both in its simplicity and in the interesting physics that stems from it. We discuss the important tests of

General Relativity that arise from the application of the Schwarzschild solution. The solution also leads us to the concept of singularities and the ever-popular black hole that has seared itself into every-day parlance. We discuss the ideas surrounding the latter.

In Newtonian physics, gravity has the curious property of making any changes in its value, its strength and direction of action, realized instantly everywhere in the universe if there is any redistribution of mass. This is usually referred to as "action at a distance". Einstein's Relativity removes this property by virtue of the character of the field equations of General Relativity. Their so-called "hyperbolic" structure allows for gravitational waves, a spreading-out through spacetime of the spacetime curvature, i.e. the gravity, at a speed of at most the speed c of light in vacuum. This in turn is in accord with the fundamental restriction imposed upon information flow in Special Relativity. Intuitively, this is very satisfactory. After all, it would not seem sensible, for example, to imagine that observers in distant galaxies, through our induced subtle local variations in gravity, could be aware of our waving our hands with absolutely zero time delay. The flow of a wave of gravity carrying the evolving information at speed c at most, is surely more intuitively acceptable than the awkward and implausible aspect of instantaneous effects demanded by Newton.

While gravity waves have been studied over the years with great interest, in this book we cover only the very basics. We also focus on an aspect of gravity waves that ties in with an issue that has been problematic in General Relativity from its early days, the issue of its energy and its localization. While the location of energy in the rest of physics is fairly straightforward, in General Relativity, the issue of its location presents new complications which have been approached in a variety of ways by various researchers over the course of nearly a century. Some of the most prominent researchers have taken the view that energy simply cannot be localized in General Relativity while others have proffered that energy, in principle, must be localizable. In this book, we focus on a particularly simple solution which is embodied in our energy localization hypothesis, that energy, including the contribution from gravity, is localized in the non-vanishing regions of the energy-momentum tensor. The latter is the very familiar and well-understood mathematical embodiment of energy and momentum from Special Relativity so our hypothesis is simplicity incarnate. We present reasons for our favouring this hypothesis and we discuss what may be viewed as a troubling consequence of the localization hypothesis, that waves of gravity would not convey energy in the course of their propagation through the vacuum.

In Chap. 7, we begin our exploration of the universe according to General Relativity. We build our picture of the basic sizes of astronomical bodies in the universe by the method of scaling; we imagine reducing the size of the Sun to that of the head of a pin and then asking on this basis, how far away the planets as well as the nearest stars are located. The process gives us a feeling for how the matter is distributed in the cosmos, something that we cannot achieve simply by having the actual incredibly enormous numbers of size and distance presented to us. We have no intuition for such numbers as they are outside of our sphere of experience.

In that vein, and following the positive experience of relying upon spacetime diagrams to help us appreciate the otherwise-confusing aspects of Special Relativity, we introduce spacetime diagrams for the evolution of simplified universe models in General Relativity. This is a subject with scope for much further development.

Having reached this point, we focus upon a subject of particular interest to the authors, the motions of the stars in the galaxies. According to Newtonian gravity, the velocities of the stars, as tracked with increasing distance from the axis of rotation of a galaxy, should be falling off in speed. Instead, the speeds have been found to be generally fairly constant. This has puzzled astronomers and physicists alike as they have believed (and most do continue to believe) that Newtonian gravity is an adequate tool for the studies of such stellar motions. After all, the motions are generally very slow by Relativity standards and the gravitational fields are generally weak, the criteria generally regarded as quite sufficient for the applicability of Newtonian gravity. As a result, researchers had concluded that vast stores of extra matter, now designated as "dark matter", must be present in huge halos surrounding galaxies in order to provide the added gravitational boost to the stellar velocities. This currently envisaged dark matter is not of the normal non-luminous matter of our every-day experience but is rather believed to interact only gravitationally and possibly also by the so-called "weak-interaction" process. In spite of intensive searches, no direct evidence for the existence of this elusive dark matter has ever materialized. Nevertheless, probably the majority of researchers regard the existence of dark matter as a given.

The present authors took a different approach to the subject. We noted that the motion of bodies that are moving solely under the influence of gravity, often referred to as "free-fall", is a more complicated subject in General Relativity than has been generally realized. As a result, we undertook a study of the modeling of a galaxy in the form of an axially symmetric steadily rotating distribution of dust. The choice of dust derived from the need to incorporate the free-fall aspect, dust being a pressure-free fluid. We developed the Einstein equations to the lowest significant order for weak fields and we were surprised to find that these equations were non-linear in contradiction to the prevailing wisdom. This was a spur to us for further investigation. We fit the known data of stellar velocities for three galaxies and our own, the Milky Way, to satisfy the field equations. This enabled us to determine the density distributions and hence the masses for the four cases. As well as fitting the so-called "rotation curves", the plots of stellar velocities as a function of their distances from the galactic rotation axis, we found that in each case, the galaxy masses were lower than those predicted on the basis of Newtonian gravity. There was no need to invoke dark matter to realize the motions. Very recently, we extended our approach to three additional galaxies with the same result. In conjunction with this study, we describe our serendipitous discovery, how Einstein's General Relativity predicts the location of the visible edge of the galaxies, now with seven cases, from the input of those mysterious rotation curves, an achievement that is unattainable with Newton's gravity. This, together with the other remarkable successes flowing from Relativity, underlines our conviction that Einstein's Relativity is truly the ultimate key to the cosmos.

In addition to developing this study, we also discuss some of the more common critical aspects that have been directed against our work. We then consider further desirable follow-up work.

In Chap. 9, we turn to the earliest analysis suggesting the need for dark matter, the anomalously high velocity of entire galaxies, taken as units comprising an ensemble, galaxies which were observed in the Coma Cluster of galaxies. Again, we confronted the possibility that General Relativity might provide a means to describe the phenomenon without dark matter. The actual physical source consists of a chaotic distribution of galaxy velocities in a cluster. Since we do not yet have a means of dealing with such chaotic systems within General Relativity, we turned to what we do have available at present, namely an idealized system of smoothly continuous matter falling radially with perfect spherical symmetry. As before, to incorporate the free-fall aspect, we consider the collapsing matter to consist of dust. In doing so, we incorporate two new aspects: exactness of solution and explicit time-dependence. Again we find that General Relativity produces results that differ considerably from those on the basis of Newtonian gravity. Most significantly, we are led to the conclusion that actual Coma Cluster data for galaxy velocities can fit this idealized model without resort to additional dark matter.

Apart from dark matter, a second different form of mysterious substance has come to be conjectured as the overwhelmingly dominant part of the universe, the so-called "dark energy". The evident need for its existence derives from the current conclusion that while the universe has undergone an expansion with the naturally expected deceleration for approximately half of its present lifetime, the latter half of the expansion has proceeded with accelerated expansion. The properly designed dark energy can be invoked to ensure that this process is followed; yet as is the case with dark matter, there is no realization of its form as an element of particle physics. As well, when we consider its connection with quantum field theory where the vacuum assumes a very active life of its own, we find an absurd requirement forced upon the theory to accord with the observations. This is discussed in Chap. 11.

In Chap. 12, we wind up our journey through Einstein's Relativity with consideration of what might be properly called "fantasy-land". Over the years, some researchers have seen in Relativity the potential for the realization of time-travel. A person takes a journey through spacetime in such a manner as to return to his or her past: Science-fiction thrillingly becoming science-fact. In this, we emphasize that the process has entailed mathematical choice of the identification of points in spacetime rather than actual physical necessity. In other words, one can make time-travel happen in the free-wheeling world of mathematics but the physical world, the world of reality, chooses it not to happen. It does so to preserve the cause-effect relationship between real events, a constraint which mathematics need not honour.

A recent popular subject is the "mutiverse", the idea that there exist universes beyond our own (even though the word "universe" has always referred to all that exists!) In our view, this is not an element of real science and however entertaining, should not be seen as anything other than fantasy. We will explore this idea and some of its byways.

We also discuss briefly the idea of spacetime worm-holes and the speculations that these have engendered.

Finally, in Chap. 13, we consider what may lie ahead as we work our way towards further understanding of our remarkable universe.

The reader will discover that there are a fair number of equations (and a few more-advanced equations) interspersed through the book. These have been placed for those readers who have more of a mathematical training than the average citizen. However the equations (as well as the appendices and the section on localized energy in Chap. 6) can be safely ignored by the more general reader who, hopefully, will still be able to achieve a solid understanding of the basic ideas without them.

Chapter 2
Preliminary Aspects of Classical Physics

2.1 Newton and His Laws of Motion

Before we get to Einstein's Relativity, we should start with Isaac Newton, who was another genius and also, in his own way, a relativist. Many, including ourselves, feel that Newton and Einstein were the greatest physics geniuses of all time. We constantly contemplate with awe what these two giants of insight and imagination have achieved, how dominant have been their respective contributions to the evolution of physics. Our focus will be on Einstein but it was Newton who most significantly set the stage for what was to come. No physics treatise can logically skip over him. Besides developing, in parallel with Leibniz, the indispensible mathematical discipline of calculus, Newton formulated three important Laws of Motion and actually, he had his own Principle of Relativity. These motion laws are very simple:

1. *A body at rest remains at rest and a body in motion, continues in motion in a straight line with constant speed unless it is subjected to an unbalanced force.*

In other words, left undisturbed, bodies keep moving uniformly in the way they have been moving. In fact, they will do so even if there are forces (pushes and pulls) on them as long as any given force is compensated for by an equal and opposite force, restoring the balance. But if there is a force on the body that does not have a balancing partner, Law 2 comes into play:

2. *If an unbalanced force acts upon a body, it is accelerated, i.e. it changes its speed and/or its direction of motion. The extent of the acceleration is proportional to the strength of the unbalanced force and the direction of the acceleration is in the direction of the unbalanced force.*

In other words, if forces on a body do not balance out, the body will *not* continue to move as it had been moving prior to being subjected to the unbalanced force. Rather, it will be accelerated, i.e. it will speed up or slow down and/or change its direction of motion, depending on which way it had been moving beforehand and which way the unbalanced force was directed. Physics students remember this law in equation form,

F. I. Cooperstock and S. Tieu, *Einstein's Relativity*,
DOI: 10.1007/978-3-642-30385-2_2, © Springer-Verlag Berlin Heidelberg 2012

$$F = ma \tag{2.1}$$

with F the unbalanced force, a the acceleration and the proportionality factor m is the mass of the body. Notice that for a given F, the bigger the m, the smaller the a. This is right in line with our everyday experience: the more massive a body, the less it will get going for a given push.

You might get confused about these two laws if you do not realize that friction is another force and it's hard to get rid of friction. But imagine a hockey puck sliding on a sheet of very smooth ice, approximating the elimination of friction. The puck just keeps going at a (nearly) steady speed in the same direction; the smoother the ice sheet, the further it goes. However, if you were to give the puck a solid smack with the hockey stick, it would speed up dramatically, and the harder the smack, the more it would speed up. Of course it does so in the direction at which it had been smacked. These laws are part of our everyday experience and make excellent sense. The Third Law is a bit more subtle:

3. *For every action, there is an equal and opposite reaction.*

If you push against the wall, the wall pushes directly back against you: action, the push, reaction, the push back. It might sound strange to think that a wall can push you but it must. The friction force on your shoes is pushing you toward the wall because you are pushing backwards on the floor (again a case of action and reaction) and since you are not moving at all, the wall must be pushing you in the opposite direction to keep the forces balanced (Law 1). So it all makes sense.

These laws plus Newton's law of "universal gravitation", that every body in the universe attracts every other body with a force F that is proportional to the product of their masses m and M and inversely proportional to the square of the distance r between them, in equation form

$$F = GmM/r^2 \tag{2.2}$$

(G is the constant of universal gravitation) are the bases of the important part of physics called "classical mechanics". It governs much of our everyday physical life-experience.

2.2 Frames of Reference

Of course the application of Newton's laws requires measurements. The measurements to which Newton's laws apply occur within frames of reference called "inertial". A frame of reference is a grid to which we can ascribe the positions of bodies at any given times.[1] In other words, it is Newton's laws that pick out inertial reference frames. A reference frame is not inertial if it is accelerated, for example a frame of

---

[1] Imagine a set of three mutually perpendicular sticks with markings at one centimeter intervals. This constitutes a reference frame. A particle's position is specified by its three-number set of position coordinates relative to this frame.

reference that is anchored to a rocket-ship while its rockets are propelling it to higher and higher speeds. Newton's laws do not hold in such a reference frame. After all, if you were to gently release a coin in such a frame, it would not stay put but rather it would move with accelerated motion in the direction of the rockets. Since there would not be any force acting on this released coin, its accelerated motion relative to the rocket-ship frame of reference would be in violation of Newton's First Law.

Newton's Principle of Relativity states that the dynamical laws of physics are the same relative to all inertial reference frames. The word "all" leads us to ask how we can generate more inertial reference frames, once we have determined a single inertial reference frame. This is easy: from one, we can create as many as we wish by simply moving with respect to the first with a constant velocity[2] (motion at a constant speed in a straight line of specified direction). The laws hold relative to these frames as well because bodies that are not subjected to unbalanced forces will still continue to move at constant velocities relative to these frames; all that happens is that each original constant velocity will just change to a new constant velocity as a result of the transformation.

In addition to the grid, we have to bring into the discussion a set of clocks to be able to determine where any given body is located as well as when it is there. For example, from a given starting point (let us say your position), the body could be 3 m to the west along the east–west line, 5 m to the north along the north–south line and 14 m up along the vertical line. Suppose it is there at a time 10, the time you set your clock to start. Those four numbers really pin it down. We would write the body's coordinates in time (the first number) and space (the second, third and fourth numbers) as a bracketed unit (10, 3, 5, 14). Physicists call this set of four numbers describing the body's position in space and time an "event" even though, as events go, this one is not really all that dramatic. It is an event in the combination of space and time we refer to as "spacetime". In our experience, it is *four-dimensional spacetime*, three dimensions of space and one dimension of time; we do not perceive any dimensions beyond these.[3] It is worth pondering for a moment why this is the case. Why do we not live in a universe with one spatial dimension? With two spatial dimensions? With four, five, six, etc. spatial dimensions? Some have even considered the notion of more than one time dimension.

If the body just stayed there relative to our grid, the first number 10 would change (ignoring smaller uncrements) to 11, 12, 13, etc. as the clock ticks away while the other numbers would remain the same. However, if the body were to be in motion relative to the grid, all of the numbers could be changing continuously as we trace out the body's history in spacetime. This grid, populated by clocks, constitutes an inertial frame of reference if Newton's laws of motion hold true relative to this frame.

---

[2] In physics, the word "velocity" takes on a very specific meaning; it has two aspects, speed that only tells you how fast the object is moving and second, direction, which way it is headed at that instant. It is indicated in physics diagrams by an arrow pointing in the motion direction with length of arrow indicating the degree of speed.

[3] Some researchers have dealt with higher-dimensional spacetimes but primarily with having the extra dimensions curled up into such an incredibly small size that they are beyond our ability to detect them.

It is worth repeating: it would not be inertial if the grid were attached to a rocket ship while the engines are blasting away. As well, it would not be inertial if it were rotating. Such frames are accelerated. Accelerated reference frames are called "non-inertial" and Newton's laws do not hold true relative to such frames. Also, it is worth repeating the example for emphasis: relative to the rocket ship frame, if you were to release a coin from rest inside the rocket ship, it would not stay put as Newton's First Law would have it but rather, it would move with acceleration in the direction to which the gases are being propelled away from the ship, just as if it had been dropped from rest as you stand on earth.[4]

2.3 Let There Be Light

We now have some of the basics that we need. The next step is to bring into the discussion the very mysterious and wondrous thing we call "light". Wondrous indeed! Imagine a universe without light. How fortunate we are to have eyes that permit us to partake of all the beauty that surrounds us and how fortunate we are for the protection that light provides for us from the many potential hazards that we face. It is even more spectacular when we probe into its essential nature. Newton had regarded light as a stream of "corpuscles", much like a hail of miniscule pellets, while Huygens claimed that light was really a wave. Curiously, in modern physics, we regard *both* Newton and Huygens as being right! Light has a dual nature, a particle aspect and a wave aspect. In modern parlance, Newton's corpuscles are called "photons". The wave aspect of light is most apparent in the many experiments that have been performed to probe its character.[5] But what do we know of waves? There are sound waves, waves that propagate in air and through water, generated by chains of molecules set into vibration. There are water waves in the seas and the oceans, there are waves traveling on a vibrating string. Common to all the familiar waves is the presence of a medium, some form of matter to "wave" the propagating disturbance. Physicists had reasoned that since we see the light reaching us from the distant stars and since the atmosphere above the earth trails off rapidly in the near emptiness of interstellar space, there must be some very rarified residual medium present to support the "waving" of light. This supporting medium must be so rarified that we cannot detect its presence in the obvious ways that we detect the media for the common waves that we mentioned. This medium was given a name: The "ether", and much of the then world of science was just as confident of its existence as it was of the existence of the traditional media, molecules, bulk water and string material respectively in the previous discussion. It is well to remember the ether when we convince ourselves that some form of matter must exist to satisfy some cherished

---

[4] In fact this phenomenon will serve as part of our lead-in to an understanding of General Relativity.

[5] Other experiments bring out the dual particle character of light. See the photo-electric effect discussed later.

theory. The theory, however seemingly inevitable, necessary and wonderful, might nevertheless be wrong. It is nature that decides.

In this regard, we must remind ourselves constantly that an essential element of any science is experimental verification. Michelson and Morley set out to detect this ether in an ingenious experiment. They reasoned that as the earth moves through the ether, an ether wind, however subtle, must be created. With the earth moving at speed v through the ether, light would have a retarded upstream speed $c - v$ as it plows its way against the ether flow and a boosted speed $c + v$ if it is directed downstream from the ether flow. This is just like a person swimming against and with a current in a river: watching the swimmer's progress from the bank, we would see the swimmer advancing more speedily in the second case than in the first.[6]

However, if light were directed perpendicular to the ether flow, it would not be retarded or boosted but just partake of the flow. Michelson and Morley arranged to have light travel in both the perpendicular and ether flow directions for equal distances and be reflected back to their origin, thus allowing the light to interfere in recombination.

Their experiment was not designed to measure the speed of light but rather to measure *differences* in light speed along different directions or paths. Under the assumption that the net speeds would differ in the different directions of travel through the ether, the light beams would constructively and destructively interfere with each other.[7] Try as they could, they were never able to detect the expected effects of the interference that would be present, had the light been traveling through an ether wind. Following this effort, Lorentz and Fitzgerald had hypothesized that bodies somehow underwent a physical contraction in the direction of motion when they traveled through the ether and hence the path length would be altered in just the right amount to remove what would have been the effects of the interference. However, many viewed this as a contrived solution to the puzzle and as we view this idea today, the Lorentz-Fitzgerald rationale appears even more naïve. In more recent times, the equivalent of the Michelson-Morley experiment has been performed using lasers, enabling far greater accuracy and the null result continued to hold.

It took the genius of Einstein to see the apparent contradiction in a totally different light (if the pun may be excused). He reasoned that there is no ether and that light has the speed c whether one measures it in any direction relative to the motion of the earth in its orbit around the Sun. Einstein's bold hypothesis: *the speed of*

---

[6] As an interesting extreme example, if the swimmer's speed relative to the river were the same as the speed of the river relative to the bank, for the observers on the bank the swimmer would appear to undergo no advancement at all when he swims upstream. His arms and legs would be flaying but he would be going nowhere relative to the observers on the river-bank!

[7] To get a simple picture of constructive and destructive interference, consider how one "interferes" with the motion of a child on a swing in the playground. To provide (maximum) "constructive" interference, that is to make the swing go higher and higher, one provides a push on the swing at the point in each cycle when the swing is at its highest point, in phase with the motion. By contrast, one renders "destructive" interference when one provides a push when the swing is still in motion towards the position of the pusher, that is out of phase with the motion. The latter has the effect of damping the swing motion.

light in vacuum is always c. The combination of Newton's Principle of Relativity, that the dynamical laws of physics are the same in all inertial frames of reference plus Einstein's hypothesis of the universality of the speed of light and extending the equivalence of inertial frames to all physical laws is generally referred to as Einstein's Principle of Relativity. The additional constancy of c hypothesis forces dramatic changes in physical theory. While it becomes somewhat more palatable to accept his hypothesis in conjunction with the idea that light can propagate in the absence of any medium, it nevertheless remains counterintuitive. After all, suppose we return to the corpuscular or quantum picture of light: little balls of energy moving at speed c. If these "quanta" (recall that they are also called "photons" when referred to light) were instead a stream of baseballs moving at speed c, then even without a medium, if we were to run toward the balls with speed v, we would measure their speed as $c + v$ relative to us and if we were to run in the opposite direction, we would measure a speed $c - v$. Relative to our catching the balls when we stood still, there would be a greater sting in our baseball glove when running towards the direction of motion of the balls and a reduced sting in the opposite direction. It is a reflection of the different energies of motion. There is a corresponding effect with light that shows itself as a shift in frequency rather than in speed of the light quanta (photons).

The truly remarkable, totally counterintuitive character of light is that it always has the same speed c in vacuum, regardless of the motion of its source or the motion of the observer. We do not see it as "speeded up" by running towards its source as we do with baseballs. Now since it is an experimental fact that science has never measured anything other than light[8] moving even at the speed c, let alone faster than c, and since we cannot boost the speed by running toward it when the speed in question is c, we conclude that according to Einstein, there is a "speed limit" for bodies and waves in nature, namely the speed c. The existence of this speed limit can also be appreciated from the fact that it would require an infinite amount of energy to bring a body of mass m to the speed of light, as we will see in Chap. 3.[9]

In turn, since all inertial reference frames are physically equivalent, that speed limit had better be the same in all the frames. It is understandable that many resisted Einstein's bold hypothesis when it was first proposed. Some of the consequences that derive from his hypothesis are even more astounding, as we shall explore in the coming chapters.

2.4 The Whole Electromagnetic Spectrum

We have discussed an essential feature of light but there is a much bigger picture to discuss. Electromagnetism is of great importance to physics in general and to Relativity as well as to our lives, of course. In what follows, we will take you through some of the basics, a very brief guided tour of the electromagnetic spectrum.

---

[8] More generally, the family of electromagnetic waves, of which light is a member.

[9] Whether this limit also applies to the speed at which information can be propagated becomes more involved in the strange world of quantum mechanics, as we will discuss later.

Light is just a small slice of the very broad electromagnetic spectrum with elements that you have heard about many times: radio waves, microwaves, infrared rays, ultraviolet rays, X-rays, and gamma-rays. The waves that allow us to see with our eyes, the light rays, fit into this list between the infrared and the ultraviolet rays. All electromagnetic waves are in the same family, composed in the same manner of oscillating electric and magnetic fields. They all travel at speed c in vacuum. The members are distinguished from each other by their wavelengths with the radio waves having the longest wavelengths (and therefore the lowest frequencies),[10] and gamma-rays having the shortest wavelengths (and the highest frequencies).

While electromagnetic waves reach us from the Sun, the stars, the distant galaxies, the quasars, the pulsars and from the deepest recesses of space, we can (and do) produce them by our own actions and our modern technologies. At the simplest level of our everyday experience, we produce them when we strike a match. At the more fundamental level, we produce an electromagnetic wave when we wiggle (i.e. accelerate) a charge. Accelerated charges produce electromagnetic waves. Imagine the wave propagating outward from the accelerating charge at the speed of light. As we wiggle it faster and faster, we observe some interesting results.

Suppose for example, we were able to wiggle the charge at a frequency of 530–1700 kHz,[11] the wavelength would be up to 600 m long and we would produce noise on the AM radio. A typical AM radio has the following numbers on the dial: 530–1700 kHz. Now, if we were to shake the charge with a greater amplitude (while maintaining the same frequency), we would produce a louder sound on the radio. This process is called Amplitude Modulation or AM. We do not have any mechanical device to wiggle a charge that quickly. Instead, radio waves are typically produced by sending charges up and down a very tall antenna (100–200 m tall).

At a rate of oscillation of 88–108 MHz, we would produce noise on the FM radio. In FM broadcasting, the loudness of the sound is not controlled by the amplitude but by the frequency. For example, if we increase our "wiggle frequency" from 99.900 to 99.915 MHz, we would progress from a quiet sound to a loud sound. This varying of frequency is called frequency modulation or FM.

At 900 and 2400 MHz we enter the spectrum of cordless phones and "wifi" internet connection respectively. Cellphone signals reside at 800,850 and 1900 MHz. VHF(very high frequency) signals have frequencies in the range 30–300 MHz and UHF (ultra-high frequency) signals span the range 300–3000 MHz for television broadcasting.

Microwaves reside at $3.10^{11} - 10^{13}$ Hz or 300–30000 GHz. Producing microwaves by sending electrons up and down an antenna at such incredibly high frequencies is not physically possible. Instead, microwaves are generated by sending electrons around a warped circle within a vacuum tube called a magnetron. One can think of

---

[10] Frequency, or cycles per second, with symbol ν is inversely proportional to wavelength with symbol λ and the product of the two is c, or in equation form, $\lambda\nu = c$. It can be seen as follows: λ is the advance in distance per cycle and ν is the cycle advance per second, hence the cycles cancel in the product leaving distance advance per second or velocity which is c.

[11] A kilohertz, kHz, is a thousand cycles (or wiggles) per second. A megahertz, MHz, is a million cycles per second. A gigahertz, GHz, is a billion cycles per second.

it as a particle accelerator. As the electrons weave through cavity, their oscillations emit microwaves.

Proceeding to higher frequencies, at the bottom end of the optical spectrum, in the region of 4.10^{14} Hz, infrared and light of red colour are next to each other. At the higher end of the optical spectrum in the region of 8.10^{14} Hz reside the colours blue, violet and the invisible ultraviolet rays. The energy quanta of each ultraviolet photon is much, much greater than, for example, the microwave photons. One can see the dramatic difference of effects produced by different types of photons in the "photo-electric" effect: if we shine red light (or microwaves) onto a metallic surface in a vacuum, no electrons would be ejected, no matter how many lamps we use. However, if we were to shine ultraviolet light onto the surface, no matter how dim, electrons would be ejected. This is a prime example of a quantum phenomenon.[12] At still higher frequencies, $10^{17}-10^{20}$ Hz, we encounter X-rays. Beyond X-rays in the spectrum, we have gamma-rays at $10^{20}-10^{24}$ Hz. We can only produce them using particle accelerators or observe them directly through nuclear decay in radioactive materials or as cosmic rays arriving on Earth from outer space.

There may be yet more powerful radiation but the most energetic phenomena that we have detected in the universe are gamma-ray bursts. These were discovered by military satellites in 1967 during the Cold War in the course of detecting nuclear weapons testing. Gamma-ray bursts are so powerful that if one were to occur within our galaxy, it could cause mass extinction on Earth. Fortunately, it is estimated that their frequency of occurrence within our galaxy is as low as one per million years and even if one were to occur within our galaxy, it is unlikely to be pointing directly at the Earth. Therefore we would not advise you to ask your insurance agent for gamma-ray-burst insurance.

---

[12] Einstein was awarded his one and only Nobel Prize for his explanation of the photoelectric effect. Many have argued that he should have been awarded at least three Nobel Prizes.

Chapter 3
A Simple Approach to Einstein's Special Relativity

3.1 Navigating in Spacetime

Your everyday experience is governed by the "where" and the "when". If you arrange to have lunch with your friend, you might tell him, "Meet me at Joe's Diner at noon tomorrow." In your mind, you know the location of Joe's Diner in the two-dimensional grid of your street map, let us say the corner of 53rd Street and 6th Avenue and if the diner is on the third floor, you know the height that you have to climb in the vertical direction. He knows it and you know it. And your watch tells you when to be there—the same for your friend. We are back to those four numbers again; we have the event of meeting the friend, where and when to do so. A presupposition is that there is no ambiguity about clocks, that there is a universal time, an absolute time (assuming that we all use good watches). We will see that Relativity changes all that.

The first author was fortunate to have spent a year in Dublin at the Institute for Advanced Studies under the directorship of J. L. Synge. Synge's heritage was remarkable. His uncle, J. M. Synge, was one of Ireland's greatest writers and nephew J. L., following in the family tradition of distinguished achievement, was a man of both letters and science. Of particular interest for us here is Synge's emphasis on the use of spacetime diagrams. By means of a graph, a picture emerges in the mind that words alone or banks of data alone cannot equal. We know this from our own experience. If we were to present lengthy tables of data giving the price of General Motors stock over a fifty year period, it would not have the same impact as the simple graph that dramatically illustrates in a single glance, the rise and fall of this corporation, as shown in Fig. 3.1. Here we have one dimension, the stock price, plotted against the other dimension, time.[1] It is even more helpful in Relativity where the mathematics can easily obscure the essentials.

Bondi [2] was influenced by Synge and he used spacetime diagrams to illustrate how one connects the observations of various "inertial observers", observers who are

---

[1] A continuation of this plot through to 2012 would reveal at a glance the significant recent recovery of General Motors.

F. I. Cooperstock and S. Tieu, *Einstein's Relativity*,
DOI: 10.1007/978-3-642-30385-2_3, © Springer-Verlag Berlin Heidelberg 2012

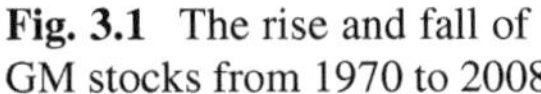

Fig. 3.1 The rise and fall of
GM stocks from 1970 to 2008

not in a state of acceleration. We will use Bondi's method and complete an element
missing in Bondi's work for the resolution of the famous "twin" or "clock paradox".

Our world is one of three spatial dimensions and time, which for mathematical
purposes is another dimension. Ideally, to illustrate our four-dimensional world, three
space dimensions plus one time dimension, we would draw four-dimensional graphs
and if we lived in five dimensions, we could do so. However, since we inhabit four-
dimensional spacetime with only three (macroscopic) spatial dimensions, at best we
can only draw in three dimensions and even this is difficult to portray suggestively on
a two-dimensional sheet.[2] Therefore, we restrict ourselves to plots in two dimensions
with only one space dimension that we call x and, analogous to how we plotted the
General Motors stock price, we use the second (space) dimension on the page to
represent the one time dimension t. In this manner, we capture the essentials of the
spacetime view. The other two space dimensions y and z are suppressed. It is a
one-dimensional spatial world view, not the complete real world but useful for our
purposes. Thus, observers move back in forth on a single common x line and we will
imagine that they are beaming light to each other along this line.

We now get back to Ms Bright, whose first name is Beatrice. We first met her in the
fantasy-rhyme in the Introduction. Actually Beatrice ("B") is one of a set of triplets
and has sisters Alicia ("A") and Camela ("C"). All three Bright women are inertial
observers. We draw our plot from the viewpoint of Alicia who is always at rest at
position $x = 0$. The vertical t axis marks off the time as read by Alicia. Therefore the
path that she follows in spacetime is the vertical line, labeled A. Suppose Beatrice,
with path labeled B, moves at constant speed v relative to Alicia and kisses Alicia
"hello" and "goodbye" while both Alicia and Beatrice have their clocks read 0,
i.e. when they cross paths for an instant. This is at the origin O with spacetime
coordinates $(t, x) = (0, 0)$ in the two-dimensional spacetime plot, Fig. 3.2. Since
Beatrice is moving at a steady speed, she advances in x proportionally to the time

---

[2] As an example of how we do so, see Fig. 5.2. The guidelines below the Sun-planet picture are
used to give a perception of depth.

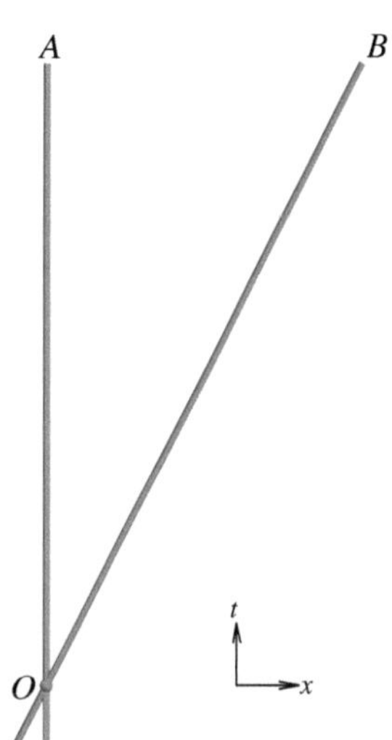

Fig. 3.2 The paths of Alicia and Beatrice in spacetime. The picture is drawn from the perspective of Alicia who does not advance in position x. She only advances in time, the vertical axis. Beatrice advances in both x and t from Alicia's perspective

that elapses. Thus Beatrice's trajectory in spacetime is the straight line at a slope relative to the vertical that is Alicia's path in spacetime. If Beatrice were to move faster, her trajectory would be a line tilted further away from the t axis, i.e. greater advance in x for a given amount of time t.

Now we bring light into the picture. Since light with speed c has the highest possible speed of propagation, its trajectory plot has the largest possible inclination with respect to the vertical t axis.[3] Suppose Alicia turns on her flashlight beaming it at Beatrice for a time period T. Clearly Beatrice receives the light for an amount of time that is proportional to the amount of time that Alicia has beamed at her; the longer Alicia sends, the longer Beatrice receives the light. At this point, we do not know how the proportionality factor relates to the speed with which Beatrice is pulling away from Alicia so, for now we will just give it a symbol k. It is called the "relativistic Doppler factor".[4]

So Alicia transmits for a period T, Beatrice receives for a proportional period kT. Now there is nothing special about Alicia as compared to Beatrice; they are both inertial observers. The physics is identical for both of them. Therefore, were Beatrice

---

[3] To determine the degree of inclination, we note that in a time t, light advances a distance x equal to the speed times the time, i.e. $x = ct$. Now relativists like to use units in which c is taken to be 1. It simplifies calculations and c is easily restored at the end if so-desired. Thus, with $x = ct$, we have $x/ct = 1$ or $x/t = 1$ using $c = 1$, the advance in x equals the advance in t for light. As a consequence, for our plots, all of the light rays are seen to be at $45°$ with respect to the t axis.

[4] The Doppler effect, the change in measured frequency as a result of source or observer motion, is familiar in our every-day life experience. We hear the steeply rising pitch from the whistle of a train and the siren of an emergency vehicle as they approach closer towards our position followed by the marked drop in pitch as they move away from us. The pitch is dependent upon the velocity of the source.

Fig. 3.3 Alicia shines light to Beatrice for a period T and Beatrice receives for a period kT and vice-versa. Note that the slopes of the light trajectories are more inclined away from the vertical axis than that of Beatrice's trajectory because light always travels faster than humans. Note that in this, as in the figures to follow, to avoid clutter, we will draw only the first and last rays of a beam transmitted over a period of time

Fig. 3.4 Alicia beams light to Beatrice for a period T and receives light back from Beatrice for a period k^2T. Beatrice receives light for a period kT and sends light back to Alicia for the same period kT

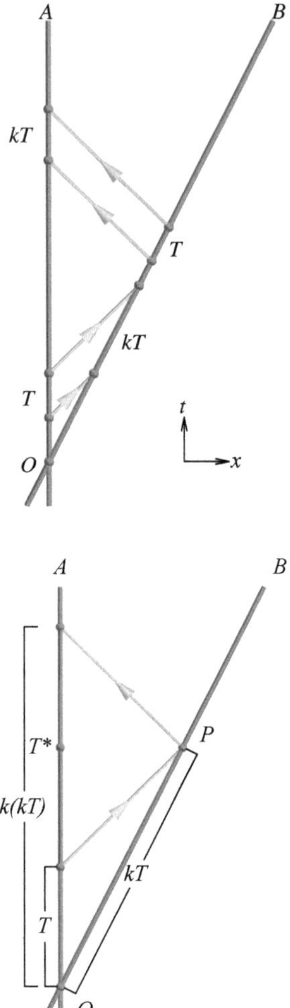

to shine her flashlight to Alicia for a period T according to her (Beatrice's) clock, Alicia must also receive the light for a period kT (Fig. 3.3).

Now let us do an experiment that allows Alicia to determine Beatrice's speed relative to herself. As soon as they cross paths at O, let Alicia begin beaming her flashlight towards Beatrice and shut it off when she reads the time T. Let Beatrice also beam her flashlight back towards Alicia beginning at O and continue to do so for as long as she receives light from Alicia. Thus, Alicia beams for a period T by her clock and Beatrice receives the light for a period kT by her (Beatrice's) clock. Beatrice is beaming back to Alicia for a period kT so Alicia receives the return beam from Beatrice for a period k times the emission period or $k(kT) = k^2T$ (Fig. 3.4).

Now Alicia is very clever and studious, reading physics books all day, so she comes up with a plan to determine that unknown k factor with the knowledge of how fast Beatrice is traveling relative to herself. She reasons that since the motion of Beatrice is uniform, Beatrice must have been at the position P in Fig. 3.4 when she was sending her last ray of light to Alicia at a time T^*. This is in the Fig. 3.4, half way between the time that Alicia sent her last ray to Beatrice and the time the last ray came back to herself (Alicia), in other words, at the time $(k^2 T + T)/2$. This is how long it took for Beatrice to reach that distance by Alicia's reckoning.[5]

If Alicia could now deduce how far away Beatrice was when she reached that point when the last ray was sent, she (Alicia) could deduce an expression for Beatrice's speed relative to her in terms of k because speed $=$ distance/time. Easily. Alicia knows that the light from herself to Beatrice travels at speed c and most importantly from Einstein, she knows that Beatrice's light beamed back to herself also travels at speed c. The ray that Alicia sent to Beatrice that reached Beatrice when she was at P was sent off at time T and the return beam arrived at Alicia at time $k^2 T$. Thus, the roundtrip time of the last light ray travel is $k^2 T - T$ and the roundtrip distance is c times this time, or $c(k^2 T - T)$. One half of this is the distance that Alicia had to know. Alicia now takes one-half of this roundtrip distance and divides it by $(k^2 T + T)/2$, the time it took to achieve that separation, to get the speed v in terms of k. The factors of $1/2$ cancel and we have for v/c:

$$v/c = \frac{(k^2 T - T)}{(k^2 T + T)} = \frac{(k^2 - 1)}{(k^2 + 1)} \tag{3.1}$$

with T cancelling to yield the final expression. If the reader is up-to-date with his or her algebra, he or she could readily solve for k in terms of v.[6] The result is

$$k = \sqrt{\frac{1 + v/c}{1 - v/c}}. \tag{3.2}$$

This k value determines how time periods get shifted when a source emits to a receiver in motion.[7]

Now if we are dealing with a wave phenomenon such as light, the period is the number of seconds per cycle of the wave and if we multiply this by the speed of the wave, namely c, we get the length of a cycle, i.e. the wavelength of the

---

[5] To see this in a step-by-step manner, note that the elapsed time OT^* in Fig. 3.4 is equal to T plus half the time between Alicia's last emitted ray and the arrival of the last ray from Beatrice as it is received by Alicia, i.e. $OT^* = T + (k(kT) - T)/2 = T + k^2 T/2 - T/2 = T/2 + k^2 T/2 = (k^2 T + T)/2$.

[6] If you would rather leave the algebra to others but still wish to gain confidence that (3.2) is the correct result following from (3.1), substitute $k = 3$ into (3.1). You will find that $v/c = 4/5$. Now substitute $v/c = 4/5$ into (3.2) and you will find that $k = 3$. Consistency! You might wish to try this with different numerical choices of k. While this process is not a proof, it certainly will give you confidence that (3.2) is the correct result.

[7] If the speed v is much less than c, we could expand the expression in (3.2) to get the much simpler approximate value $k = v/c$.

received light relative to that of the emitted light. Notice that if v is positive as in the Fig. 3.4 with Beatrice pulling away from Alicia, (3.2) shows that k is always greater than 1 (with v a positive value),[8] and therefore the period and hence the observed wavelength is always greater than the emitted wavelength. This is the well-known and very important phenomenon of "red-shift", the shift of the wavelength of the waves towards the red end of the spectrum of colours in the rainbow.[9] As well, from our cosmological observations of redshift, we know that the universe is in a state of expansion with the distant galaxies rushing away from us. It is also noteworthy that (3.2) is an exact expression, holding true even if Beatrice's speed relative to Alicia is nearly as great as the speed of light, c. Of course it can never reach c, *not* for material objects like people! We will soon see why.

Instead of having Beatrice move away from Alicia with speed v, we could have arranged it for her to be moving towards Alicia, i.e. with speed $-v$. In that case, the derivation would have gone on as before and the net result would have been the same as (3.2) but with v replaced by $-v$. With this replacement, the new expression shows that k is less than 1, the period kT observed by Beatrice is now shorter than the emitted period and the light is seen as blue-shifted rather than red-shifted, i.e. shifted towards the shorter wavelength blue end of the visible spectrum. If astronomers had seen the distant galaxies blue-shifted, they would have deduced that the universe is in a state of contraction with the galaxies rushing towards each other and that we were possibly heading towards the "Big crunch".

Notice that when we replaced v by $-v$ everywhere in (3.2), the result is the same as if we had taken (3.2) and flipped it upside down. In other words, letting v go into $-v$ leads to k going into $1/k$. This will be useful in what follows later.

Finally, Camela is getting into the action. Suppose she also travels relative to Alicia but with a different constant speed. Let v_{AC} be the speed of Camela relative to Alicia, let v_{AB} be the speed of Beatrice relative to Alicia that we had before and let v_{BC} be the speed of Camela relative to Beatrice. You trust from your everyday experience that if you are watching a train go by at 100 km/h and people in the train are moving toward the front of the train at 5 km/h, they appear as if they are moving at 105 km/h relative to you, watching from the ground. In other words, the speeds appear to simply add up. In the case of the triplets, we would have simply

$$v_{AC} = v_{AB} + v_{BC}. \tag{3.3}$$

Surprisingly, this is not quite correct in Relativity but you would not notice the mistake unless the speeds were much higher. To arrive at the correct expression, we have to distinguish between the k factors that relate the triplets' observations.

---

[8] Note that if $v = 0$, $k = 1$ and the periods are the same. This makes sense because in this case, Beatrice and Alicia are at rest relative to each other.

[9] The colours of the rainbow can be remembered by their first letters in order as "roygbiv", red, orange, yellow, green, blue, indigo and violet. (If you pronounce it a few times as if it were a word, it sticks in your mind forever!) This is the visible spectrum with the red colour having the longest wavelength and the violet colour having the shortest wavelength in the visible part of the electromagnetic spectrum.

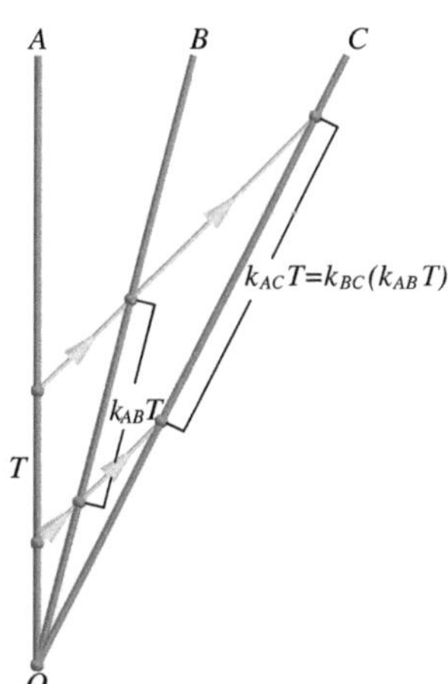

Fig. 3.5 Alicia beams light to Beatrice who in turn beams light to Camela for the same period that she receives light. Because of the invariance of the speed of light, this two-step process follows the same lines as the one step of Alicia beaming light directly to Camela

Let k_{AB} be the k factor relating Beatrice to Alicia that we had before, let k_{AC} be the k factor relating Camela to Alicia and let k_{BC} be the k factor relating Camela to Beatrice. Let Alicia shine light to Beatrice for a period T. Beatrice receives the light for a period $k_{AB}T$ and as long as she receives this light, she shines light to Camela. Camela receives Beatrice's light for a period equal to Camela's k factor relative to Beatrice times Beatrice's emission period, in other words $k_{BC}[k_{AB}T]$ as in the Fig. 3.5.

The very interesting thing is this: Thanks to Einstein, we could have just as well seen this two-step process as the single-step emission of light from Alicia to Camela. The lines would have been identical because the speed of light is always the same c relative to any pairs of participants in the experiment. Viewed as the Alicia-to-Camela emission, Alicia sends for a period T and Camela receives for a period $k_{AC}T$. Now the reception period for Camela is the same, whether she received the light in the two-step process or directly from Alicia so we can equate the two values. Canceling the common factor T, we have

$$k_{AC} = k_{AB}k_{BC}. \tag{3.4}$$

If the triplets' mother Delilah (labeled D) decided that she wished to participate in the same way, the compounding of the k factors would become $k_{AD} = k_{AB}k_{BC}k_{CD}$ and so on if we wish to include cousins and other relatives. We see that the compounding of k factors in Einstein's Relativity is just as simple as the compounding of velocities in classical non-relativistic physics. It is simply a matter of multiplication of k factors here replacing addition of velocities in classical pre-Relativity physics.

This raises the question: what is the relativistic replacement for the classical compounding of velocities (3.3)? For this, we apply (3.1) and (3.4):

$$v_{AC} = c\frac{k_{AC}^2 - 1}{k_{AC}^2 + 1} = c\frac{(k_{AB}^2 k_{BC}^2) - 1}{k_{AB}^2 k_{BC}^2 + 1} \tag{3.5}$$

and we use (3.2) in (3.5) to change the expression on the right hand side from one that is in terms of k_{AB} and k_{BC} to the final expression in terms of v_{AB} and v_{BC}. After a little bit of simple algebra, this gives the relativistic law for the compounding of velocities,

$$v_{AC} = \frac{v_{AB} + v_{BC}}{1 + v_{AB}v_{BC}/c^2}. \tag{3.6}$$

Complexity is now evident relative to the simplicity of compounding in classical physics. Clearly if the relative velocities are small compared to c, the correct relativistic expression will differ only slightly in comparison to the classical expression of (3.3). However, if the velocities are appreciably close to c, the results of compounding will be dramatically different. As a concrete example, suppose v_{AB} and v_{BC} are both $7c/8$. The classical compounding law (3.3) would give a value of $7c/4$ for v_{AC}, comfortably reaching well beyond c. However, you can easily verify that the correct result for v_{AC} using (3.6) is $112c/113$, very close to c but not quite c! You might wish to try to exceed c with v_{AB} and v_{BC} still closer to c but that denominator will never let you reach c with compounding, let alone pass it. Nature's speed limit cannot be breached![10]

3.2 The Relativity of "Now"

In terms of the spacetime diagrams, let us now see how different observers view pairs of events differently. Let us return to Alicia and Beatrice and consider two explosion events L and R (see Fig. 3.6) that Alicia deduces to have occurred equidistant from her at position $-x$ and x respectively. She knows that the explosion sites were equidistant from her because at time $t - x/c$, she sends out flashes to her left and to her right to arrive at the events L and R and they return to her simultaneously at time $t + x/c$. She also deduces from these observations that the events L and R must have both occurred simultaneously at the half-way time between the emission time and the reception time of the returned flash, namely at the time t.

Beatrice meets Alicia at the time t. She also goes through the same steps as Alicia. She sends out her flash at time t^*_{Rbef} to reach the event R and she receives the return flash at time t^*_{Raft} as shown in Fig. 3.6. Similarly, she send a flash at time t^*_{Lbef} to arrive at the event L and she receives the return flash at time t^*_{Laft}. Like Alicia, Beatrice also deduces that the flashes occurred equidistant from her because she notes that the time intervals for the round trips of her flashes to R and L, namely $t^*_{Raft} - t^*_{Rbef}$ and $t^*_{Laft} - t^*_{Lbef}$ respectively, are precisely the same. Actually we can see this clearly from the symmetrical positioning in Fig. 3.6.

---

[10] But see later discussion in Chap. 13.

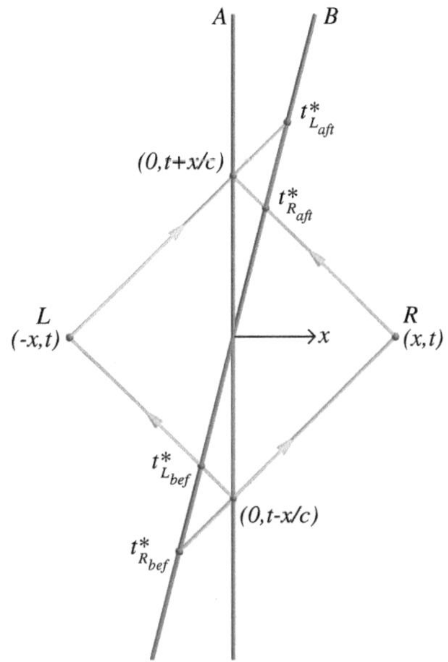

Fig. 3.6 Alicia and Beatrice each send beams of light that arrive at sites of explosions to their left and right. For Alicia, the explosions occur simultaneously but for Beatrice, the explosion to her right occurs prior to the explosion to her left

But the interesting thing is this: Beatrice sent out her flash to *R before* she sent out her flash to *L* and she received the return flash from *R before* she received the return flash from *L*. Beatrice deduces that event *R* happened *before* event *L*.[11] The important deduction: *Events simultaneous in one frame are not simultaneous in another frame.*

It is important to note that the blending of the observations for the two observers would not proceed in this simple manner were it not for the essential fact upon which the edifice of Relativity is constructed—that light rays from the observers who are in relative motion actually move in unison over the indicated segments. As Bondi has described it, "There is no overtaking of light by light". This would not be the case for a Newtonian observer.

What we learn from this picture is that the "now" of one observer will differ from the "now" of another observer moving relative to the first in Relativity. This phenomenon has led some researchers, such as the renowned mathematician K. Gödel, to view time itself in an entirely new light. Some have come to question whether time is eternal or whether time had a beginning and will have an eventual end with the birth and death of the universe respectively (presupposing that the universe is of finite lifetime). What are we to make of these speculations? Do they have any tangible consequences for physics or are they only of relevance to philosophers? Does it really make any sense to speak about time (or space, for that matter) if the universe does not even exist?

---

[11] Note that Beatrice would have deduced that *R* happened after *L* if she had been moving in the $-x$ direction rather than the $+x$ direction relative to Alicia.

3.3 Relativity Transformations for Space and Time

Let us return to our grid and our set of clocks telling us where something is present or something happens, and when it is there or when it happens. It is important to be able to connect the "where" and the "when" as seen by one observer, say Alicia, in terms of how another observer, say Beatrice, designates the "where" and the "when", i.e. the transformation of the coordinates in spacetime of an event. We can derive the essentials by keeping with our picture of Beatrice moving with speed v relative to Alicia along their common x axes.

Let Alicia refer to distance along this axis as x and let Beatrice refer to distance along this axis relative to her as x^*. In restricting ourselves to this picture, we are setting the other two spatial dimensions relative to Alicia (call them y and z) and Beatrice (call them y^* and z^*) to have the same corresponding values for all time. If Alicia and Beatrice set their clocks to 0 when they cross at O as in Fig. 3.7, Beatrice will be at position $x = vt$ at time t relative to Alicia and Alicia will be at position $x^* = -vt^*$ at time t^* relative to Beatrice. For a Newtonian, time is absolute and there is no distinction between the clock readings t and t^*; for Alicia and Beatrice, according to Newton, $t = t^*$. However, we will see that in Einstein's Relativity, time is no longer an absolute.

We return to our spacetime plot for Alicia and Beatrice and now consider an event E that occurs at position x at the time t as reckoned by Alicia and at position x^* at the time t^* as reckoned by Beatrice.[12] Alicia and Beatrice collaborate in an experiment by having a ray of light go out from each of them in such a manner as to arrive precisely at the position and time of the event E, and then be reflected back to each of them. Referring to Fig. 3.7: let Q and Q^* be the corresponding points of the return ray. The ray must leave Alicia at time $t - x/c$ by her reckoning since the light travels a distance ct in the time t. In this manner, the ray arrives at event E precisely at time t in Alicia's frame. Now Beatrice must do likewise (but in terms of her coordinate system), have a ray leave her at a time $t^* - x^*/c$ which insures that her ray will also arrive at the event E at position x^* relative to her at the time t^*. Now by Einstein, the speed of the ray for Alicia is the same as the speed for Beatrice; it is c for each of them. As a result, the ray from Beatrice lies coincident with the ray from Alicia and we can apply the same relations between the intervals that we used before. We only use one k factor so let us define $k = k_{AB}$. From before, we have the interval OP^* related to the interval OP as

$$(t^* - x^*/c) = k(t - x/c). \tag{3.7}$$

The ray "illuminates" the event E and we let it reflect back to Beatrice first and then to Alicia as shown. The reflected ray reaches Beatrice at the time of reflection plus the time to return the distance x^*, i.e. at the time $t^* + x^*/c$. Similarly, it gets back to Alicia at time $t + x/c$ relative to her clock. To find the relationship between these

---

[12] We do not prejudice the issue and let Beatrice have a clock rate of her own. If it should work out to be the same as that of Alicia's clock, so be it. We shall see!

intervals, we note that here, it is Beatrice emitting to Alicia rather than vice versa. Since the speed of now Alicia relative to Beatrice is to be taken into account, the velocity to be used is $-v$. We recall that when v goes into $-v$, k goes into $1/k$. As a result, the OQ reception interval is related to the OQ^* emission interval as

$$(t^* + x^*/c) = \frac{1}{k}(t + x/c). \tag{3.8}$$

We leave it as an exercise to use (3.7) and (3.8) and to eliminate k with (3.2) to get

$$x^* = \frac{(x - vt)}{\sqrt{1 - v^2/c^2}}, \quad ct^* = \frac{(ct - vx/c)}{\sqrt{1 - v^2/c^2}}. \tag{3.9}$$

Note that (3.9) gives the values of x^* and ct^* in terms of x and ct. However, suppose we wish to know the values of x and ct in terms of x^* and ct^*. If you are nimble with your algebra, you can manipulate the equations in (3.9) to get

$$x = \frac{(x^* + vt^*)}{\sqrt{1 - v^2/c^2}}, \quad ct = \frac{(ct^* + vx^*/c)}{\sqrt{1 - v^2/c^2}}. \tag{3.10}$$

However there is an easy way to arrive at this inverse result: since Beatrice has velocity v relative to Alicia, Alicia has velocity $-v$ relative to Beatrice. Also, for the inversion of (3.9), we need Alicia to assume the role of Beatrice and Beatrice to assume the role of Alicia. This is achieved by having the un-starred coordinates of Alicia go into the starred coordinates of Beatrice and vice-versa everywhere in (3.9). By the simple actions of these velocity and coordinate switches in (3.9), we see that (3.10) is the result (note that $(-v)^2 = v^2$). A little bit of thinking eliminated the algebra work!

The factor $1/\sqrt{1 - v^2/c^2}$ appears very frequently in Special Relativity, so it is helpful to define it as γ. It is a scaling factor which determines how much Relativity stretches or contracts a dimension. At low velocity, say, $v = 1$ m/s, γ is very close to 1 but at velocities comparable to the speed of light, say 99 % of the speed of light, $\gamma = 7.089$. For a still higher value, at a velocity of 99.99 % of the speed of light, $\gamma = 70.712$. You can see how dramatically the effect grows with increasing velocity. Equation (3.9) is referred to as the "Lorentz transformation". It finds its way into many applications, some of which we will discuss in this book. The immediate point to note from the second of (3.9) is that t and t^* are no longer the same. *Time is no longer an absolute as it was in pre-Relativity physics.* This is one of the most amazing and to many, one of the most unsettling aspects of Einstein's Relativity. There is an essentially new mixing between x and t in the Lorentz transformation. This mixing of space and time will be discussed in the next section.

As we have seen before, the distinction becomes more and more essential for relative velocities between Alicia and Beatrice getting closer and closer to that of light. The difference fades into insignificance for velocities v much less than c which is the case for our common daily life experience. That is why you were able to

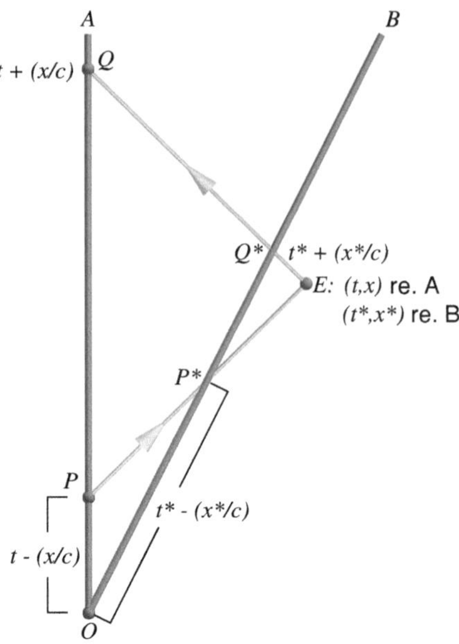

Fig. 3.7 The event E coincides with the arrival of the rays from both Alicia and Beatrice. The connection between intervals from Bondi's k-calculus is used to build the transformation between the space and time coordinates used by Alicia and Beatrice

make a reliable arrangement to meet your friend at Joe's Diner at a time that you both agreed upon without any problem, without invoking Einstein. However, suppose you were "relativistic" observers, observers moving relative to each other with speeds approaching c. Then the situation would become most interesting indeed, as we will now witness.

3.4 The Twin or "Clock" Paradox

The twin paradox is one that has captured the imagination of many ever since it was first discussed at the dawn of Relativity. We now delve into it with the help of our delightful Bright sisters. Twins Alicia and Beatrice (Camela will have a role soon as well) synchronize their clocks and Beatrice sets off on a long journey in her rocket ship at a speed nearing the speed of light relative to stay-at-home Alicia. At a certain point, Beatrice decides to come home, reverses direction and eventually returns to the home of her twin Alicia. She discovers that there are strangers living in the house, wearing clothing that she had never seen before and using electronic devices that she could never have imagined possible. After discussions with the inhabitants, Beatrice is shocked to learn that she is speaking to the great-great grand-children of her beloved late sister.

The paradox consists in imagining that Beatrice had stayed put and it was really Alicia who had made the roundtrip. Then one might have believed that it should be Alicia who should be conversing with Beatrice's great-great grand-children.

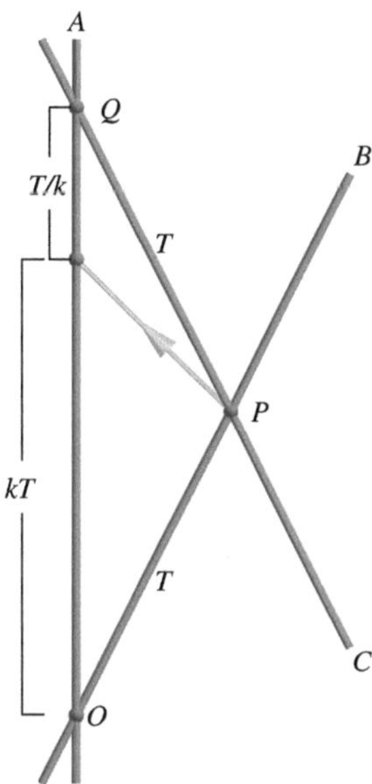

Fig. 3.8 Beatrice and Camela use their synchronized clock readings to simulate a twin who makes a return voyage from home. The traveling twin returns, aged less than the stay-at-home twin

"After all", one might say, "motion is relative". Not really in this case. While inertial reference frames are totally equivalent physically (when Alicia and Beatrice had a constant velocity with respect to each other in our earlier examples), both Alicia and Beatrice could not have been inertial at every stage in this case. Beatrice is the one who makes the return voyage in an *absolute* sense in that she was the one who underwent a period of deceleration followed by acceleration. She had to employ her rocket thrusters and during the periods of deceleration and acceleration, even if the rockets were very silent and her space-ship acoustically sealed, she could feel the effects of being in her now non-inertial reference frame during the crucial periods. She could feel it in the forces on her body as she sits strapped into her seat. There is an essential asymmetry between Alicia's and Beatrice's journeys and it is this asymmetry which leads to Beatrice rather than Alicia who returns to the world of her sister's great-great grand-children rather than vice-versa. We will now see why it is Beatrice, the rocket ship adventurer, who is the one to live to witness the strange experience.

Returning to Alicia and Beatrice, let Beatrice and Alicia synchronize their clocks at the intersection point O and let Beatrice beam light to Alicia immediately after their meeting at O for a period T (Fig. 3.8). Precisely when Beatrice's clock strikes T, Beatrice meets with sister Camela, who this time happens to be traveling towards Alicia with the same speed that Beatrice is moving away from Alicia. Camela, like Alicia and Beatrice, has an excellent clock and Camela sets her clock to time T at the instant she meets Beatrice (point P in the diagram).

Camela begins to beam light to Alicia at that point and she continues to do so until she meets Alicia at point Q in the diagram. Clearly from the symmetry of equal and opposite velocity, Camela's beaming period is T just as it was for Beatrice. Since Camela's time was T when she met Beatrice at P, it is $T + T = 2T$ when she meets Alicia at Q. The question is: what time does Alicia read at Q?

To determine this, we note that since Beatrice is beaming to Alicia for a period T, Alicia receives the light for a period kT where k is the relativistic Doppler factor between Beatrice and Alicia. We also recall that with v changing into $-v$ as is the case for Camela's velocity relative to Alicia, the relativistic Doppler factor between them becomes $1/k$. Therefore, with Camela beaming to Alicia for a period T, Alicia receives Camela's beam for a period T/k. We deduce that the time elapsed for Alicia between O, her meeting with Beatrice, and Q, her meeting with Camela, is $(kT + T/k) = (k + 1/k)T$. While Camela says it is $2T$ o'clock at Q, Alicia says it is $(k + 1/k)T$ o'clock.[13] You can easily satisfy yourself that $k + 1/k$ is greater than 2 for all positive values of k (except for $k = 1$ which is of no interest since for this special case there is no motion at all), and so we conclude that Alicia has aged more than her triplet sisters. Notice that it is the asymmetry that determines which way the relative aging goes: the single inertial observer Alicia ages more than the combined inertial observers Beatrice and Camela. There was one observer in the first case, two observers in the second case.

Our natural inclination would be to think of the combination of Beatrice and Camela as being equivalent to the earlier described case of Beatrice alone, blasting off in a rocket ship and returning younger than Alicia when she returns home. This is because we could imagine the Beatrice-Camela combination being almost like a Beatrice voyage alone, with the only difference being a short deceleration/acceleration period turnaround for Beatrice. However, the following objection is immediately raised by various critics: while the different results hold for the readings of the clocks when all the observers are inertial as described in Fig. 3.8, it does not follow that this will be equivalent to replacing the two inertial observers Beatrice and Camela with the single *noninertial* observer Beatrice. In fact various authors as well as pundits have argued that the clocks of Alicia and Beatrice will get back into synchronization during the deceleration/acceleration period, and Alicia and Beatrice will reunite with the same time reading at Q.

In this regard the first author recalls the wonderful lectures that Bondi gave at the Brandeis Summer Institute in Theoretical Physics in 1964 when he raised this very issue. In his discussion, Bondi suggested that with the deceleration and acceleration applied very gradually, there should be no significant effect on the clock, but he admitted that the issue remained unresolved. Through the years, some have argued that smooth as this may be, nevertheless the accumulated effect must be one of a re-adjustment of the clocks to have their readings coincide at Q.

The counter-argument that we have advanced is as follows [1]: consider a variety of journeys of different durations by Beatrice with the same constant velocity but with all of the journeys having identical deceleration/acceleration turnaround periods. We have illustrated three of such journeys in Fig. 3.9. The essential point is this: the different journeys result in time differences for Alicia and Beatrice that vary from

---

[13] It is a very simple algebra exercise to show that $k + 1/k = 2\gamma$.

Fig. 3.9 Three different
return voyages for Beatrice
with each having identical
turnaround characteristics.
Since they are identical, they
could at most re-adjust Beatrice's clock by exactly the
same amount in the three
cases. However the three voyages would require three different adjustments to get back
into synchronism with Alicia's clock. Thus, we see that
the deceleration/acceleration
effects could not bring the
clocks of Alicia and Beatrice
into synchronization

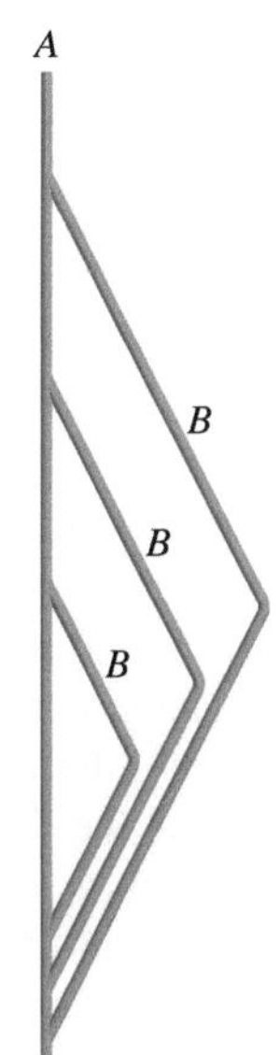

journey to journey as a result of the different values of T. Therefore, the decelera-
tion/acceleration periods would have to induce *different* degrees of make-up in time
for the different journeys to get the times to agree for the various cases. However,
assuming a turnaround does force a make-up in time, it must do so by precisely the
same amount in all the cases, because the turnaround is identical and there has been
no change in the state of the underlying spacetime for all of the journeys.[14] Thus we
see that the argument that the deceleration/acceleration phases provide the required
adjustment periods is inconsistent with the *varied* make-up requirements and the
conclusion of asymmetric ageing holds.

The result is certainly counter-intuitive, so it is worth considering this very interesting phenomenon in different ways. The essential aspect that Relativity brings to
bear here concerns the nature of time, that it no longer enjoys the absolute status that
it had in classical physics. Various treatments of Relativity refer to time as being just
another coordinate, on the same basis as the coordinates of three-dimensional space.
However this is very misleading. It is true of course that Relativity dethrones time
as an absolute, in contrast to position, which, even in classical physics, is dependent
upon the observer.[15] But time's new status in Relativity is *not* one of equivalence
to space; rather it bears a *reciprocal* relationship to space. Its essential distinction
from that of space remains in Relativity but it combines with space in a very special

---

[14] We will have more to say about the nature of evolving spacetimes in General Relativity. Here
we are dealing with Special Relativity where the spacetime is always the same.

[15] Example: if A is moving relative to B, A is always at the position O, the origin of A's coordinate
system anchored on her body whereas B records a continuously different position of A as time
advances.

manner that we will discuss in what follows. Einstein's Relativity leads us to deal with this wonderful marriage of concepts that we call "spacetime" in a manner that essentially changes our view of physics.

3.5 Perceptions of Length and Time

Let Alicia hold a "rigid" meter rod firmly in her grasp. As Beatrice moves with velocity v relative to Alicia along the x axis, Beatrice wishes to determine the length of the rod as she sees it in her frame of reference. Suppose the left end is at position x_1 and the right end is at x_2 in Alicia's reference frame. Since it is at rest in Alicia's frame, the ends have these values for all times t of Alicia's clock. The 1 m of the rod is the difference between the right-end coordinate and the left-end coordinate, i.e. $1\,\mathrm{m} = x_2 - x_1$. In physics, we call the length that is read in the rest frame of an object, its "proper length". Thus Alicia's length of 1 m that she measures is the rod's proper length.

We wish to find the length of the rod that Beatrice perceives. We use (3.10), first for the right end position x_2 and then for the left end position x_1:

$$x_2 = \frac{(x_2^* + vt_2^*)}{\sqrt{1 - v^2/c^2}}, \quad x_1 = \frac{(x_1^* + vt_1^*)}{\sqrt{1 - v^2/c^2}}. \tag{3.11}$$

Now we have to consider an issue that we might easily take for granted. It is well to ask: which times t_2^* and t_1^* are we to choose? Actually it does not matter *as long as they are the same*. This is the only logical way to define the length as Beatrice sees it: it is the difference of the end coordinates $x_2^* - x_1^*$ with each end coordinate *taken at a common time $t^* = t_2^* = t_1^*$*. For Beatrice, her length is this "snapshot" of the end-point differences.

Thus with $t_2^* = t_1^*$, we subtract the second equation from the first in (3.11), the t_2^* and t_1^* terms cancel each other, and we find

$$x_2 - x_1 = \frac{(x_2^* - x_1^*)}{\sqrt{1 - v^2/c^2}} = \gamma(x_2^* - x_1^*) \tag{3.12}$$

or

$$l = l^*/\sqrt{1 - v^2/c^2} \tag{3.13}$$

where the lengths for Alicia and Beatrice are respectively

$$l = x_2 - x_1, \quad l^* = x *_2 - x_1^*. \tag{3.14}$$

Note from (3.13) that since $\sqrt{1 - v^2/c^2} = 1/\gamma$ is always less than 1 for velocities v that are not 0, we have the result that l^* is less than l. Beatrice perceives Alicia's meter rod to be less than 1 m in length. Clearly Alicia's perception of length, the proper length, is the maximum length that can be measured for the rod. Measurement relative to a frame having any motion relative to the rod brings in the $1/\gamma$ factor which is less than 1 for any v different from 0. *Proper length is maximal.*

At this point, you might suspect that we are trying to sell you the Brooklyn Bridge. You might wish to raise the objection that objects moving relative to you do not have the appearance of having shrunk. However, your experience with motions of objects are such that their velocities are much smaller than the speed of light, c. Suppose you were to witness an object moving by at the incredible speed of 10 km/s. You can easily calculate that for this velocity, $1/\gamma = 0.999999995$! Even with this extraordinary velocity, the effect of Relativity is so very minute. However in particle accelerators, with particles approaching the speed of light, the relativistic effect is appreciable and it cannot be neglected. This Relativity effect is witnessed on a daily basis in labs throughout the world.

We now return to the issue of the perception of time. We considered this in the previous section but now we will examine it using the Lorentz transformation. Let Beatrice carry her clock which is at her position x_B^*. It is *always* at this position relative to her coordinate system because she is carrying it. Let Beatrice record two ticks of the clock, tick 1 at time t_1^* and tick 2 at time t_2^*. Consider the times to which these ticks correspond in Alicia's reference frame. We use (3.10) for the two ticks in turn:

$$ct_1 = \frac{(ct_1^* + vx_B^*/c)}{\sqrt{1 - v^2/c^2}}, \quad ct_2 = \frac{(ct_2^* + vx_B^*/c)}{\sqrt{1 - v^2/c^2}}. \tag{3.15}$$

We subtract the first equation from the second, the x_B^* terms cancel and we are left with

$$c(t_2 - t_1) = \frac{c(t_2^* - t_1^*)}{\sqrt{1 - v^2/c^2}}. \tag{3.16}$$

Thus we see that the time interval $T = t_2 - t_1$ as read in Alicia's frame is related to the interval $T^* = t_2^* - t_1^*$ as read by Beatrice's clock as

$$T^* = T/\gamma. \tag{3.17}$$

The relativistic γ factor appears as before but with a different significance. The time T^* read in the rest frame of the clock (Beatrice's frame) is called the "proper time". Note from (3.17) that Alicia's time T is necessarily longer than Beatrice's proper time T^*. *Proper time is minimal; proper length is maximal.* This *reciprocal* relationship between the proper quantities again underlines the essential difference between space and time in Relativity. Time is *not* just another coordinate like space.

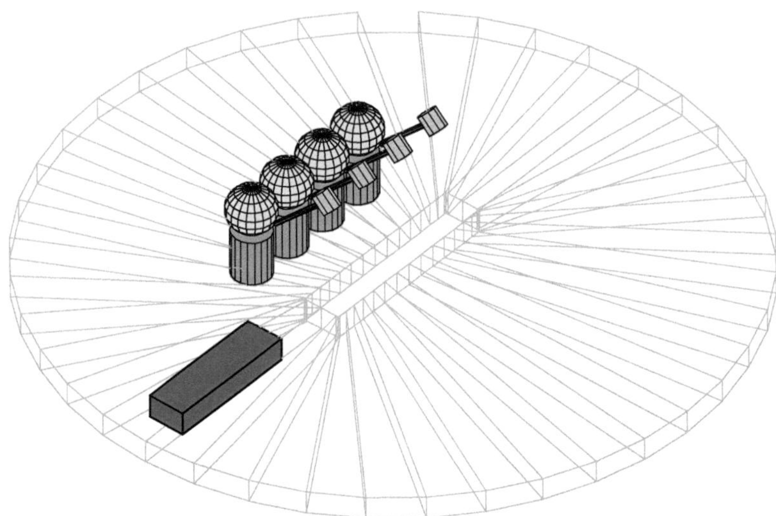

Fig. 3.10 The plank, seen as shortened by Alicia and her friends, approaches the channel. They prepare to strike the plank simultaneously by their reckoning

3.6 More Paradoxes in Special Relativity

Earlier, we dealt with the interesting twin paradox and now we turn our attention to two paradoxes involving lengths. Suppose a channel that is 10 m long and 1 m wide is carved out of the ice and Alicia and a group of her friends are standing with hammers along its length. Beatrice invites a group of her friends to stand with her along her 10 m "rigid" plank and they go hurtling along the ice at a relativistic velocity (Fig. 3.10). By Relativity, Alicia and company view Beatrice's plank as somewhat shortened and when it is in a position over the channel, Alicia and her friends simultaneously take solid smacks at the plank with their hammers (taking care to strike only in the spaces between the plank riders!) (Fig. 3.11). The plank descends through the ice channel and into the icy water below (Fig. 3.12).

However, examining the situation from the vantage point of Beatrice and her friends, they see the channel that has shrunk. After all, relative to them, it is the channel that is in relativistic motion while their plank is at rest and it is always the moving object that appears shorter. The paradox that presents itself is clear: How can this now-shortened channel accommodate the longer plank from the viewpoint of Beatrice and company?

To resolve the paradox, two important aspects come into consideration: simultaneity and rigidity. While the smacks are simultaneous for the members of the Alicia troupe, a careful use of (3.10) reveals that they are *not* simultaneous for Beatrice and her friends who are standing on the plank. Beatrice et al see the leading end being hit first and the following points along the plank hit in a *succession* of blows along its length. The leading end enters the water first and succeeding plank portions

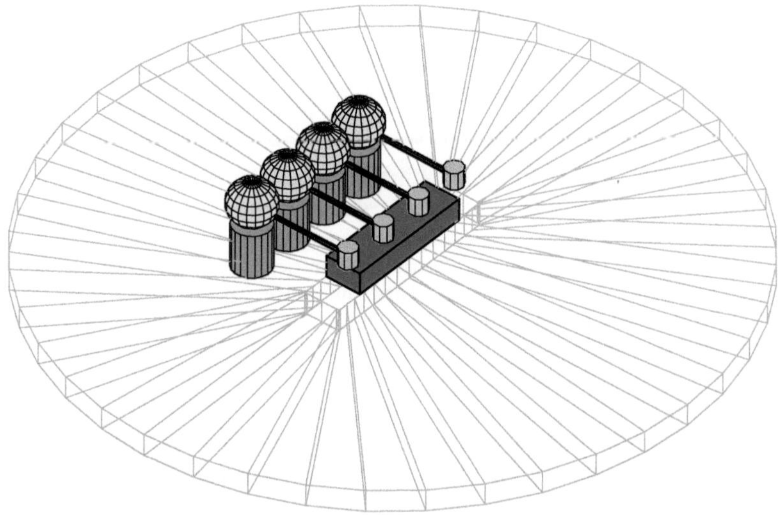

Fig. 3.11 Alicia and her friends smack the plank simultaneously while it is totally over the channel

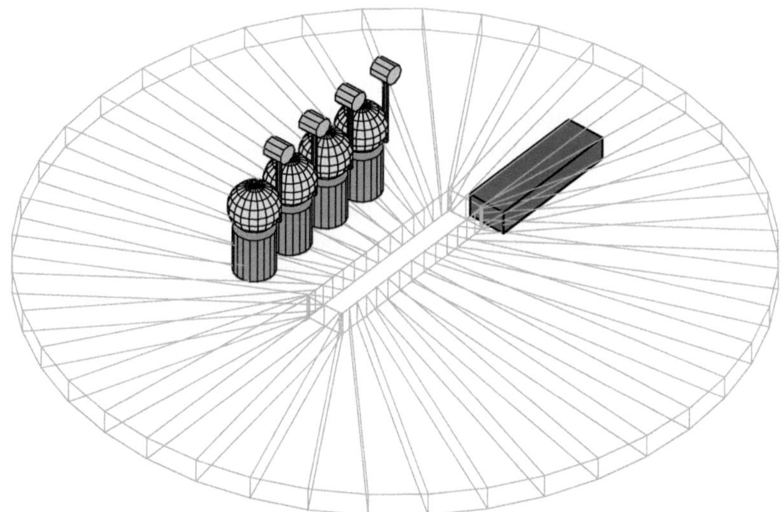

Fig. 3.12 The plank is now submerged under the ice and continues its forward motion

follow in turn as the plank seems to slither into the water below as if it were a noodle
(Fig. 3.13)!

However the plank was supposedly rigid, yet here we are speaking of a flexible
noodle. This demands an explanation and the one that we face might seem somewhat
astounding: *truly rigid bodies cannot exist in Relativity*! It is easy to see why this
is so. Suppose the plank in question were truly perfectly rigid. If that were true,

Fig. 3.13 Seen from the vantage point of Beatrice and her friends, it is the channel that is shortened. They see the Alicia crew smack the leading end first followed by successive smacks along the length of the plank and the plank enters the ice in a noodle-like shape. After all the elements of the plank have submerged, the straightened plank continues its forward motion under the ice

all of its elements would always have to move in unison as there could never be any buckling. However, suppose the plank were to be smacked at one end. As soon as it was smacked, that end would move. Since it is supposedly rigid, every portion of the plank would have to move *simultaneously* with the smacked end. To do so, the pulse from the smack would have to be transmitted along the plank with infinite speed, but we know that effects can propagate with at most speed c. Thus the plank must buckle, however minutely. The moral of the story: simultaneity is relative and absolute rigidity does not exist.

Some years ago, we posed this paradox on a test to a group of students, asking them to present the resolution. One answer stood out from all the rest. The resourceful student, let's call him Willy Wiseguy, reasoned as follows: because of Relativity, the results are different for the Alicia and Beatrice observers. The Alicia observers discover that the Beatrice group, along with the plank, end up submerged in the water below whereas the Beatrice observers, seeing a shortened channel, never actually pass through the ice and into the water but simply continue on their way along the ice! Hmm. Very interesting, Willy. But your grade is an "F" nevertheless. At the end of the operation, the plank is either above the ice or below, within the water. It is either wet or dry. It cannot be both.

Another interesting paradox concerns the rod and the barn doors. Alicia has her two friends stand guard at the front and rear doors facing each other 20 m apart in the family barn. The front door is open. Beatrice holds a rod whose proper length in its natural uncompressed state (recall: this is the length that is measured in the rest reference frame of the rod) is slightly longer than 20 m. Alicia is determined to have the rod fit into the barn so she directs Beatrice to run so fast that Alicia sees the rod appear to have shrunk to 20 m. By this means, the rod can fit into the barn (Fig. 3.14).

Alicia has directed her friend to snap the front door shut when the rod ends are precisely at the front and rear doors. The rod is in the barn (Fig. 3.15). The next step follows: the rod is now trapped but in a compressed state. It is compressed because it has a shorter length than it had when it was measured in its rest frame before being trapped. As it decompresses and expands, it blasts the rear and front doors open as the rod expands back to its original uncompressed proper length (Fig. 3.16).

Fig. 3.14 In the figure, the rod is seen approaching the barn from the viewpoint of the rest-frame of the barn. From this viewpoint, the rod which is longer in the rod rest-frame, is seen, in the barn rest frame, to have a sufficiently shortened length so that it can fit completely into the barn

Fig. 3.15 As seen from the viewpoint of the barn rest-frame, the front door of the barn is shut when the rod is totally within the barn

Now let us consider the situation from Beatrice's viewpoint. For her, the barn appears to have shrunk so that the distance between the front and rear doors is even less than 20 m yet Beatrice is running with her longer-than-20-m rod, as she perceives its proper length (Fig. 3.17). One would have to decide: either the rod did manage to get inside the barn or it did not get inside the barn. We can well imagine Willy's response: "it's relative. The rod gets into the barn as far as Alicia is concerned but it never gets in according to Beatrice."

Again, poor Willy falls short. However it must be said that this example is somewhat more complicated since we are now dealing with changes in velocity in the direction of initial motion.

For Beatrice, the barn appears to be even shorter than 20 m. According to her view, the leading end of the rod has stopped at the rear door while the rest of the elements of the rod are still in motion with a portion of the rod still outside the front door (Fig. 3.17). As time progresses for Beatrice, the elements of the rod make their way through the front door until the rod's trailing end reaches the front door (Fig. 3.18). At that point, the front door is shut, an event *later* than the event of the front end of the rod reaching the rear door. Then, the stresses build up to the point where the doors

Fig. 3.16 From the viewpoint of the barn rest-frame, the trapped rod blasts the rear and front doors open as the rod expands back to its proper length

Fig. 3.17 The rod is pictured as it approaches the barn from the viewpoint of the rest-frame of the rod. In this frame, the barn appears to be even shorter than 20 m

cannot constrain the rod and it pops out the doors from Beatrice's vantage point as well (Fig. 3.19). For Beatrice, the stresses build in a different sequence from the one experienced by Alicia.

The common theme in these as in other paradoxes revolves around the issue of the Relativity of simultaneity. We can imagine how difficult it must have been even for scientists in the days of Einstein's first formulation of Special Relativity, as it is not comfortably accepted by many, even to this day.

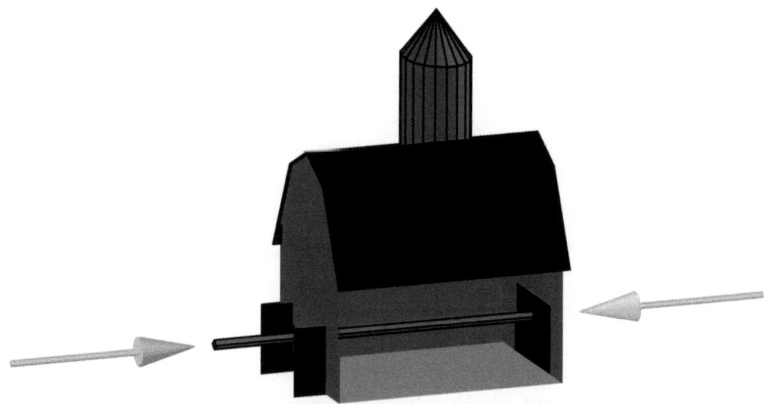

Fig. 3.18 From the viewpoint of the rest-frame of the rod, the leading end gets stopped at the rear door while the rod elements are compressing to the point where the trailing end of the rod enters the front door

Fig. 3.19 From the viewpoint of the rest-frame of the rod, the rear door gets blasted open slightly before the front door gets blasted open

3.7 Exploring Spacetime

We are familiar with our feelings about the relativity of time in a psychological sense. For example, we might say to a friend "Doesn't it feel just like yesterday that we were on that hiking trek in the mountains?" However we know that in actual fact, it was three months ago. In earlier sections, we began to come to terms with the powerful new idea that in Relativity, time is relative in a very real sense. It is not just in our minds. We are now ready to probe further and familiarize ourselves with the important concept of "spacetime", Relativity's amalgam of space and time. To really appreciate how this comes together, we will have to use some mathematics, but it will be kept simple.

We start with the beautiful theorem of Pythagoras that we learned in grade school. If we have a right-angle triangle with side lengths x and y, the hypotenuse length l is related to the side lengths as in Fig. 3.20,

$$l^2 = x^2 + y^2. \tag{3.18}$$

The simple extension of this theorem to a rectangular box of sides x, y and z with distance l between opposite corners is (Fig. 3.21)[16]

$$l^2 = x^2 + y^2 + z^2. \tag{3.19}$$

We now wish to make the leap from space to spacetime. Alicia and Beatrice shift into 3-dimensional space and Alicia, whose reference frame coordinates are (x, y, z), arranges to have a flash of light shine from an emission time t_1 at position (x_1, y_1, z_1) and be absorbed at time t_2 at position (x_2, y_2, z_2) in her frame of reference. She could station observer 1 at (x_1, y_1, z_1) and observer 2 at (x_2, y_2, z_2), each with clocks and each at rest relative to Alicia. Their life experience is the same as that of Alicia; they are part of her frame of reference. In terms of Alicia's coordinates, the distance l_{12} that the flash of light travels, using (3.19) is given (as a square) by

$$l_{12}^2 = x_{12}^2 + y_{12}^2 + z_{12}^2 \tag{3.20}$$

where $x_{12} = x_2 - x_1$, $y_{12} = y_2 - y_1$, $z_{12} = z_2 - z_1$. However, there is another way to express the distance that the flash has traveled. After all, distance equals velocity-times-time and since the velocity is c and the time interval is $t_{12} = t_2 - t_1$, we could just as well write the distance as

$$l_{12} = ct_{12}. \tag{3.21}$$

We now can equate the square of the distance expression in (3.21) to its value as expressed in (3.20) and if we take the latter to the other side of the equation, we are left with 0:

$$c^2 t_{12}{}^2 - x_{12}{}^2 - y_{12}{}^2 - z_{12}{}^2 = 0. \tag{3.22}$$

Now Beatrice, who is moving with velocity v relative to Alicia, first considers the event of the emitted flash. For her, it has for the *spacetime* coordinates, the time[17] that it was emitted and the place from where it was emitted in her starred frame of reference, $(ct_1^*, x_1^*, y_1^*, z_1^*)$. The event of flash absorption according to Beatrice is $(ct_2^*, x_2^*, y_2^*, z_2^*)$. In the same manner as we had for the events in the Alicia reference frame, we have for Beatrice's reference frame

---

[16] We could continue on in this vein with boxes of 4, 5, 6 etc. spatial dimensions but we stop at 3 because our (at least macroscopic) spatial world is 3-dimensional. Note that we cannot draw the pictures for 4 and higher spatial dimensions.

[17] It is handier to multiply the time by c.

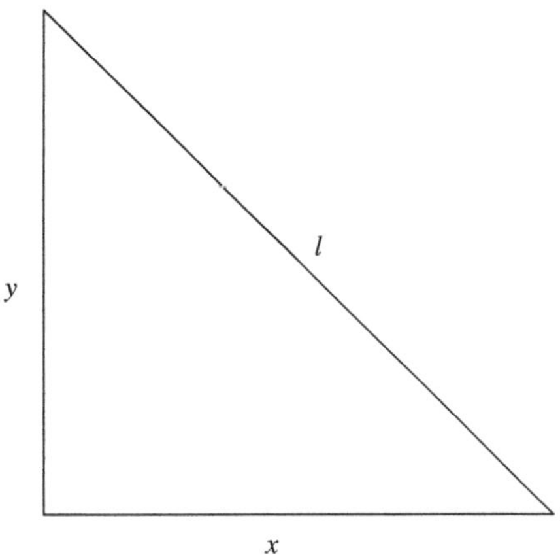

Fig. 3.20 The sides x, y of a right-angled triangle are related in length to the hypotenuse l as $l^2 = x^2 + y^2$

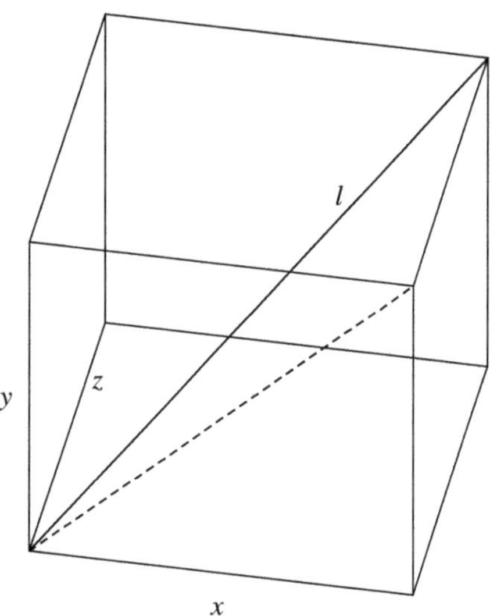

Fig. 3.21 The sides x, y, z of a rectangular box are related to the diagonal l as $l^2 = x^2 + y^2 + z^2$. The *dotted line* indicates the hypotenuse of Fig. 3.20

$$c^2 t_{12}^{*2} - x_{12}^{*2} - y_{12}^{*2} - z_{12}^{*2} = 0, \tag{3.23}$$

in other words everything that we had for Alicia, only now everything is starred *except c*. The reason that c is unstarred is simple and important: $c = c^*$, the invariance of the speed of light.

The expressions on the left hand sides of (3.22) and (3.23) are of particular significance in Relativity. They embody the square of what we call the "spacetime interval" s_{12} (and s_{12}^* relative to Beatrice) between the two events 1 and 2, in this case, the events on the path of a light ray:

$$s_{12}^2 = c^2 t_{12}^2 - x_{12}^2 - y_{12}^2 - z_{12}^2. \tag{3.24}$$

and

$$s_{12}^{*2} = c^2 t_{12}^{*2} - x_{12}^{*2} - y_{12}^{*2} - z_{12}^{*2} \tag{3.25}$$

in Beatrice's starred reference frame. From (3.22) and (3.23), we see that for the events on the path of a light ray, both s_{12} and s_{12}^* are zero. Now suppose those events 1 and 2 were very very close together. In calculus, we write the infinitesimally small difference between the spacetime coordinate values and the spacetime interval between the two events as

$$dx = x_2 - x_1, \quad dy = y_2 - y_1, \quad dz = z_2 - z_1, \quad cdt = c(t_2 - t_1), \quad ds = s_2 - s_1 \tag{3.26}$$

which allows us to express (3.24) as

$$ds^2 = c^2 dt^2 - dx^2 - dy^2 - dz^2 \tag{3.27}$$

and (3.25) as

$$ds^{*2} = c^2 dt^{*2} - dx^{*2} - dy^{*2} - dz^{*2}. \tag{3.28}$$

Earlier we saw that when the finite spacetime interval was zero for Alicia, it was also zero for Beatrice. If it is so for finite quantities, it is also true for the infinitesimal quantities ds and ds^*. Now suppose that we are dealing with a pair of events that do *not* lie on the path of a light ray. Such pairs of events cannot be separated in such a manner as to produce a zero spacetime interval because whatever the relationship between them, it is not mediated by a signal at the speed of light c. So we now consider the connection between ds and ds^* in general. In Appendix A, we provide a simple proof that

$$ds = ds^* \tag{3.29}$$

in general, even when they are not zero.

The important result that emerges is that the spacetime interval between any two given events is an invariant.

It tells us how the special combination of space and time is something upon which everyone must agree. Its structure is also of interest: notice that it is the *positive* time interval squared and the *negative* of the sum of the space intervals squared that combine to give us the invariant. This again points to the essential distinction between time and space. Time is not just another coordinate like the space coordinates. The difference in the sign in front of the quantities spells this out clearly.

3.8 Types of Intervals and the Light Cone

We have already encountered the special type of interval connecting two events that lie on the path of a light ray, the emission of a flash and its subsequent absorption. This type of interval is called "light-like" or "null"; its length is zero, $s_{12} = 0$. The reason

Fig. 3.22 Alicia screams and some distance from her position, a goblet shatters

that we can achieve zero length with the necessarily positive values of the squares of
the distances and the square of the time interval is because the spacetime intervals
have pluses and minuses in front of them, plus in front of the time interval squared
and minuses in front of the space intervals squared. In the light-like intervals, the
pluses and minuses perfectly balance each other to give a net value of zero. Light-
like intervals represent the limiting boundary of what we term "causally connected
events", those pairs of events, 1 and 2, for which event 1 can cause event 2. In the
flash example, the event of absorption was dependent on there being the event of
emission. There was a cause-and-effect linkage.

Let us consider an example of a different type of event pair: suppose Alicia at her
position $x = 0$ at the instant $t = 0$ emits a high-frequency scream. Some distance x
from her position, a fine crystal goblet sits on the table and at the slightly later time t,
the goblet shatters into many pieces (Fig. 3.22).

The spacetime interval s squared between the events of the scream (Event $(0, 0)$)
and the shattering (Event (ct, x)) is[18]

$$s^2 = c^2(t - 0)^2 - (x - 0)^2 = c^2t^2 - x^2. \tag{3.30}$$

We now consider the sign of s^2 in (3.30). To determine the sign, we consider the physical connection between the events. The goblet shatters because Alicia has induced
high-frequency oscillations of the air molecules in her mouth and these oscillations,
the sound, has traveled through the air at the speed of sound, v_s, until they reached
the goblet and the induced vibrations sufficed to shatter it. There was the cause, the
scream, and the effect, the shattering. The sound traveled a distance x in the time

---

[18] $y = 0 = z$ for both Alicia and the goblet.

t so its speed was $v_s = x/t$. Thus $v_s^2 t^2 = x^2$. If we substitute this value for x^2 in (3.30), we have

$$s^2 = c^2 t^2 - v_s^2 t^2 = (c^2 - v_s^2)t^2. \tag{3.31}$$

Now the speed of sound is less than the speed of light so we see that s^2 is positive and therefore s is a real number. Events separated by a real number spacetime interval are said to be "timelike separated" events. Note also that the shattering of the goblet occurred *after* Alicia's scream. Now we know that Relativity can produce some surprising effects so it is well to ask whether Beatrice, moving relative to Alicia, could ever see the events occur in *reverse* order. In the footnote below, we show that this can never happen.[19] It is very well that this can never happen. It would be very disconcerting for us if it were possible for Beatrice to travel with sufficient speed that she should see the shattering *before* the scream! It would play havoc with our common sense because the scream *caused* the shattering. Without Alicia's scream, that goblet would have continued its existence intact in its original state. To sum up, the causally related events occur in a definite order in time for all observers.

It is useful to show this in a spacetime diagram bringing into our lexicon for the first time, the "light cone" (Fig. 3.23). As before, we draw the spacetime picture from Alicia's viewpoint. She is at the origin O of her (t, x) coordinate system and at time $t = 0$, she screams. The goblet shatters at a later time t_s at some distance x_s away from her. We also have rays of light flash by her from left to right and from right to left just as she screams. We draw these rays with just the right slopes so that $x/t = c$ for the left-to-right ray and $x/t = -c$ for the right-to-left ray.[20] Note the position of the shattering goblet event: it occurs at a spacetime point that lies between the vertical time axis and the left-to-right propagating light flash. It lies there because we can trace the sound wave that came from Alicia's mouth to the goblet and the speed v_s of the sound wave, traveling slower than c, follows the path that lies closer to the t axis than the light ray path.

The idea of the preservation of order in cause-effect relationships is very important. Later we will discuss how some researchers have taken seriously the notion that this

---

[19] We use the second equation of (3.9). Here v is the velocity of Beatrice relative to Alicia. For the scream event for Alicia, $x = 0$, $t = 0$ so we see from the equation that it occurs at time $t^* = 0$ as well for Beatrice. The goblet-shattering event is at time t^* given by the second of the equations (3.9) and therefore, for Beatrice, the separation in time (the later time minus the earlier time) between the events is $t^* - 0 = t^*$. If we substitute $v_s t$ for x, we have $t^* = \gamma c t (1 - \frac{v v_s}{c^2})$. Clearly this is always positive since both v and v_s are always less than c. Thus t^* is necessarily positive and Beatrice sees the shattering *after* the screaming just as Alicia did.

[20] In the figure, we have gone beyond the use of only one spatial dimension x. We have imagined that light rays flashed to Alicia from every direction in the x–y plane, arriving at her position just as she screamed. Then, instead of two single rays as in our earlier spacetime diagrams, we have two *cones* of rays, one cone of rays going in, the backward or past light cone, and one cone going out, the forward or future light cone. Ideally, we would like to draw a three-dimensional light cone since we inhabit three-dimensions of space. However, we require one dimension of the plot to mark off the passage of time and we have no dimension left to do so after using the three spatial dimensions for the illustration. We have to give up the aid of the spacetime diagram in this case and work only with the mathematics.

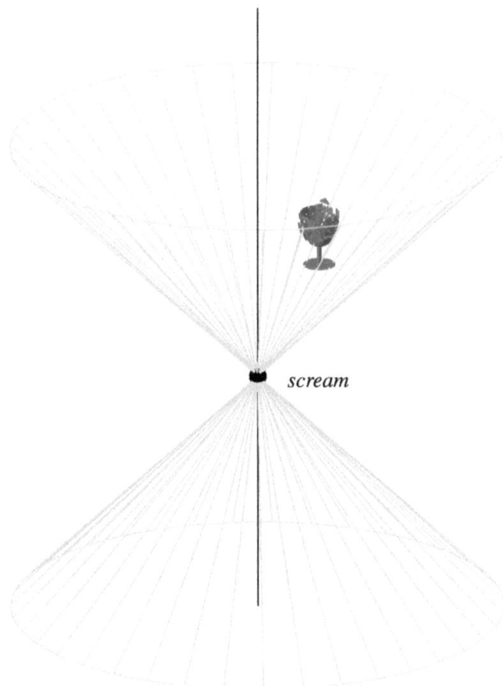

Fig. 3.23 In this figure, we see the event of the shattering of the goblet at a time later than the event of the scream. The shattering event lies within the future light-cone of the scream event

ordering can be violated in physics. The subject is connected with what are technically called "closed timelike curves" or "time machines" in the popular parlance. We will also show how we have approached the issue, concluding that they are a mathematical artifact rather than an element of physical reality.

Now let us consider a different pair of events. The first event is again Alicia's scream but now the second event is the landing of an astronaut on Mars (Fig. 3.24). Let us suppose that the landing event occurred 5 s after the scream in Alicia's reference frame but at precisely the same instant in Beatrice's reference frame. We are not at all surprised that this could be so; after all, Alicia's scream certainly has nothing to do with the Mars landing in the sense of cause and effect.[21] While we are aware that they followed in that particular order for Alicia, we would naturally say "so what?". We would also say "so what?" if Camela happened to be travelling at the right velocity to witness Alicia's scream *after* the Martian landings. The earlier transformations can be used to show these possibilities and we are not at all surprised. Without a cause-effect relationship, it does not really affect our sense of logic. However, it is important to note that in the old Newtonian way of looking at reality, there was an *absolute* time and there was an absolute ordering of events, whether they were of the first variety, the scream and the shattering goblet, or the second variety, the scream and the Mars landing. One of the events followed the other in that particular order

---

[21] This is so even if there were air between the Earth and Mars to carry the sound wave!

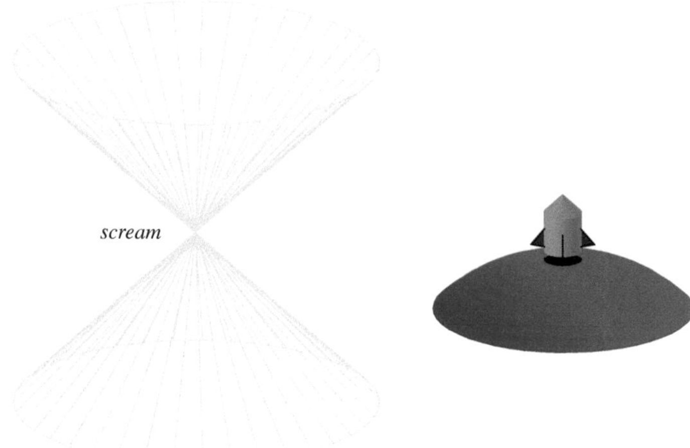

Fig. 3.24 In this case, Alicia screams and in her reference frame, 5 s later, an astronaut lands on Mars. However Beatrice determines that these two events occurred simultaneously in her reference frame. There is no contradiction as the Mars landing event occurs outside of the light cone of the scream event

for *everybody*. For Newton, time was absolute. For Einstein, time takes on a whole new meaning.

Let us probe further into the Mars landing on the spacetime plot. It occurred 5 s after Alicia's scream so if we were to draw a line to indicate a ray that connected the scream to the landing, it would have to be a *fictitious* ray, i.e. unphysical because it would have to be traveling at a speed greater than that of light.[22] After all, even light cannot travel from Earth to Mars in 5 s! On average, it would take approximately 12 min for light to travel from Earth to Mars. We say that the Mars landing occurs outside of the light cone from Alicia's scream.

The spacetime interval squared between the scream and the Mars landing, $s^2 = c^2t^2 - x^2$ is a negative number because x^2/t^2 is greater than c^2. Thus, the interval itself, the square root of a negative number, is an imaginary number. Events separated by an imaginary spacetime interval are said to be "spacelike separated" because they occur at spatially different locations for all observers. How do we know this? Easy! For suppose there existed a reference frame of someone like Beatrice for whom the events occurred at the same place in space. If that were so, then the interval squared would be $s^2 = c^2t^2 - 0$ which is a positive number. But then, the s^2 value, which, we recall is an invariant, being positive negates the fact that it must be negative for all observers including Beatrice. So we cannot have these events occur at the same spatial position for *any* observer; they are absolutely separated spatially.

So we have three kinds of regions *relative to the scream*, the $O = (0, 0)$ event in our spacetime plot: the region within the upper light cone is the "absolute future"

---

[22] See however, the discussion in Chap. 13 where it was claimed that the recent neutrino experiment reveals a transport of particles with speed exceeding c.

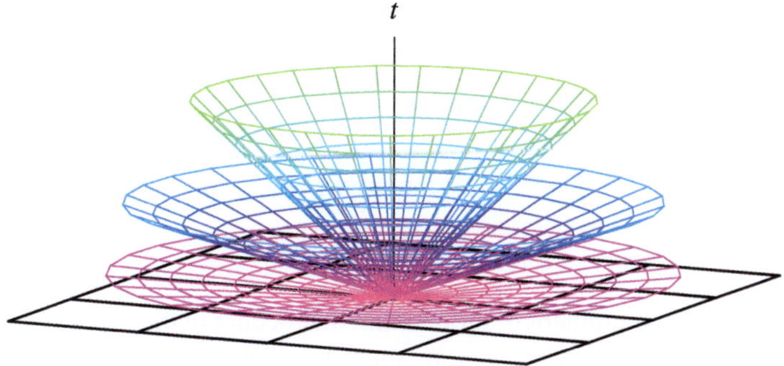

Fig. 3.25   The opening-up of the light cones as the speed c is taken to increase. For infinite c, the light cone is replaced by a plane representing the Newtonian "now". At that stage, the "before" zone is the entire region below the plane and the "after" zone is the entire region above the plane

relative to O; the region within the past light cone, with an analogous argument, is the "absolute past" relative to O and the entire region elsewhere we designate as the "absolutely separated" region relative to O. The remarkable difference in Newtonian physics is that the absolutely separated region does not exist! For Newtonian physics, which embodies interactions at infinite speed of propagation, the upper and lower cones open up to the extent that they are both coincident with the (x, y) plane. The cone structure is no longer present and we are left with simply the absolute past and the absolute future. At that stage, the (x, y) plane represents the "now" where all of the events are coincident in time with the scream event at time $t = 0$ (Fig. 3.25).

   We will find it very useful to have this understanding about events in spacetime when we come to the discussion of gravitational collapse and black holes.

3.9 Energy-Momentum—the Relativity Modifications

Let us return to our earlier discussion about the important mechanical laws of Newton. Recall from the Second Law that when an unbalanced force F acts on a body, it gets accelerated and the acceleration a is proportional to the unbalanced force. The proportionality factor, the property of the body involved, is what we call its mass, m, the measure of its inertia, if you like, the extent to which it resists any change in its existing state of rest or motion. Expressed in equation form,

$$F = ma. \tag{3.32}$$

This is one of the most important equations in classical physics. It also appeals to our common experience: for a given force F, the bigger the mass m that is being pushed, the smaller will be its acceleration. Actually the more general way to express this law, to account for the fact that the mass might be varying, by for example shedding fuel exhaust, is by means of the quantity that we call momentum, p:

$$p = mv. \tag{3.33}$$

The improved version of (3.32) is

$$F = \frac{dp}{dt}. \tag{3.34}$$

Here, dp is the little increment of momentum that occurs in the course of the little bit of time dt in which it makes this momentum increment. From (3.34), the standard development that you might wish to read about in the mechanics text books, leads to an expression for "work" which is the force acting through a distance and how this work gives the change in the energy, E of the body. The energy assumes the form

$$E = \frac{mv^2}{2}. \tag{3.35}$$

In Relativity, we follow a more general law of physics which is astonishingly successful and beautiful. It is called the "Principle of Least Action". The law is that for any physical system, there exists an invariant quantity S called the "action" such that the system, in going from one state to another state, follows the route which minimizes this action. Many physicists regard this law as the most fundamental in all of physics. An interesting simple account of this law is to be found in [4] and a more sophisticated technical approach meant for more advanced students of physics in [3]. By applying the least-action law, we find that the proper expression for momentum is no longer mv but rather

$$p = \gamma mv \tag{3.36}$$

and the energy is no longer $\frac{mv^2}{2}$ but rather

$$E = \gamma mc^2. \tag{3.37}$$

It is easy to see that for small velocities compared to the velocity of light c, the relativistic momentum expression (3.36) merges with the classical mechanics expression (3.33) since γ approaches 1 in that case. However the relativistic energy expression takes on an important change. As we see from the above equations, while the classical energy and the classical and relativistic momentum go to zero when v goes to zero, the relativistic energy goes to

$$E = mc^2. \tag{3.38}$$

Remarkably, it tells us that at its base level, matter at rest, every bit of matter, regardless of how seemingly insignificant, is endowed with an intrinsic energy by virtue of its having mass. If we put in the numbers, we can appreciate just how astounding an amount of intrinsic energy is contained in even a tiny amount of mass. As an example, consider a mere wisp of matter, let us say one gram of sugar, which hardly shows up in a teaspoon, and imagine it being converted entirely to energy. By (3.38), this tiny amount would produce 9.10^{13} Joules of energy, enough to run a thousand-watt heater for three thousand years!

Another important point to note is that γ in (3.37) would become infinite if the body could ever attain the velocity c of light. This again points to the speed limit c in nature.

Chapter 4
Introducing Einstein's General Relativity

4.1 What is Gravity?

In the preceding chapters, we witnessed the new aspects of physics brought about by Einstein's Relativity. All of this arose because: (a) there is a speed limit in nature, c, the maximum speed at which influences can be propagated and (b) because all inertial observers are physically equivalent, they must all agree on this speed limit. However, not a word was mentioned about gravity and it is with the inclusion of gravity that Einstein's Relativity takes on a whole new and exciting complexion. Einstein's Relativity without gravity is called "Special Relativity" to distinguish it from Relativity with gravity which is called "General Relativity". In brief, General Relativity is Einstein's theory of gravity. In what follows, we will delve into General Relativity, showing how it is the curving of spacetime in the general theory that replaces the old Newtonian idea of gravity being just another force.

The essence of gravity was appreciated by the cave-men and cave-women: If you were a cave-dweller and you let go of your club, you found that it fell to the ground. If you were to step off the edge of a cliff, you discovered that you would fall, and that your speed would increase more and more as you fell further and further. The further the fall, the more it would hurt you so you learned from childhood to take gravity very seriously. If you were a cave-dweller, the power of gravity was an essential part of your existence, as it is for us to this day. It was natural to think of gravity, that agency that makes objects fall, as a force like the others.

Delving further, you would notice that a bird feather fell more slowly to the ground than did a rock so you would naturally build into your mind-set the idea that lighter objects accelerate less than heavier objects in falling under gravity. However, you would be misguided because the air-resistance had been playing a crucial role in retarding the fall of the feather more than of the rock. If you were a more scientifically inclined cave-woman, you might have picked up a pebble and a rock and dropped them together from a substantial height after positioning your significant-other cave-man to act as observer. You would probably have been very surprised to learn from him that the pebble and the rock hit the ground simultaneously. In this experiment,

F. I. Cooperstock and S. Tieu, *Einstein's Relativity*,
DOI: 10.1007/978-3-642-30385-2_4, © Springer-Verlag Berlin Heidelberg 2012

the air-resistance played a much less significant role. In fact, a dramatic illustration of the fact that bodies of different masses fall with the same acceleration under gravity is achieved by re-doing the feather-and-rock experiment in a vacuum tube where almost all of the air has been removed. The feather and the rock would be seen to hit the bottom simultaneously.

Newtonian physicists of the 17th century had understood this well. After all, if we were to combine (2.1) and (5.1), we have

$$ma = GmM/r^2 \tag{4.1}$$

where we now let m be the mass of the pebble or the rock and let M be the mass of the Earth that is gravitationally attracting them. The mass m cancels on both sides of the equation and we see that the acceleration a is

$$a = GM/r^2 \tag{4.2}$$

which is *independent* of the mass m, the mass of the pebble or the rock. Thus, pebbles and rocks fall the same way.

Soon we will discuss how Einstein used this result to launch his new theory of gravity. However, we will first consider why Einstein was confronted with the necessity to formulate his new theory. It is well to ask because Newtonian gravity, in conjunction with Newton's laws of motion had served physics well for centuries in describing how the Moon revolved around the Earth, the planets revolved around the Sun and in more modern times, how rocket trajectories were determined. The immediate problem for Einstein was that issue of the speed limit in nature. In Newton's gravity, when the distribution of mass changes, for example if you were to stretch out your arms, the gravitational influence of your doing so would be felt instantaneously throughout the universe!

For the benefit of those with some background in calculus,[1] the form of Newton's law of gravity in differential form is

$$\nabla^2\phi = 4\pi G\rho \tag{4.3}$$

where $\nabla^2 = \frac{\partial^2}{\partial x^2} + \frac{\partial^2}{\partial y^2} + \frac{\partial^2}{\partial z^2}$, ϕ is the gravitational potential whose gradient determines the local acceleration and ρ is the mass density. This differential equation is mathematically of the "elliptic" type for which changes propagate instantaneously. Thus, a change in ρ in the example described above, as the removal of mass density from one location (along the sides of your body) and relocating it to a new position (perpendicular to your body), would instantaneously alter ϕ throughout the universe and the acceleration experienced at all points in the universe would change without any delay whatsoever. This would contradict Special Relativity, the theory which has been remarkably successful in experiment after experiment.

---

[1] This paragraph can be skipped by non-calculus readers.

Fig. 4.1 Standing in an elevator on the surface of the Earth, Einstein releases a pebble and a rock from rest. They fall along (nearly) parallel lines at the same rate of acceleration

It is worth repeating: The very successful theory of Special Relativity pivots on the concept of a finite invariant speed limit in nature. So Einstein set himself on the path of discovering a new theory of gravity, one which would meld properly with Special Relativity.

Various physicists including the first author's mentor, N. Rosen, who worked closely with Einstein, have remarked upon a key aspect of Einstein's approach: simplicity. Einstein treasured simplicity and had the faith that the essential truths in physics would be simple. His lead-in to a new theory of gravity bears witness to this credo. Going back to those pebbles and rocks, Einstein imagined some experiments[2] in an elevator. In Fig. 4.1, we have Einstein in a stationary elevator holding a pebble and a rock. Released from rest, the pebble and the rock fall at the same rate to the floor, as we discussed before.

Now Einstein imagined this elevator imbedded in a rocket ship far from gravitating matter. The rockets are blasting, accelerating the elevator upwards in the Fig. 4.2 at a rate of $9.8m/s^2$, the acceleration that objects experience at the surface of the Earth. Now with Einstein releasing the pebble and the rock, he experiences the same phenomenon that he did on Earth: He witnesses the pebble and the rock "fall" to the floor with an acceleration of $9.8m/s^2$, just as before on Earth. Einstein saw the two phenomena as being physically *equivalent*. However, relative to observers outside the rocket ship, the released pebble and rock appear to hang suspended (Newton's First Law) while it is Einstein who appears to be accelerating upwards at a rate of

---

[2] These are frequently referred to as *gedanken* experiments from the German, meaning "thought" experiments, experiments not actually performed but rather imagined to have been performed. Einstein was particularly fond of such exercises. He employed them in his intellectual battles with N. Bohr over the issue of the probabilistic interpretation in quantum mechanics. See [5] for an interesting account of these exchanges.

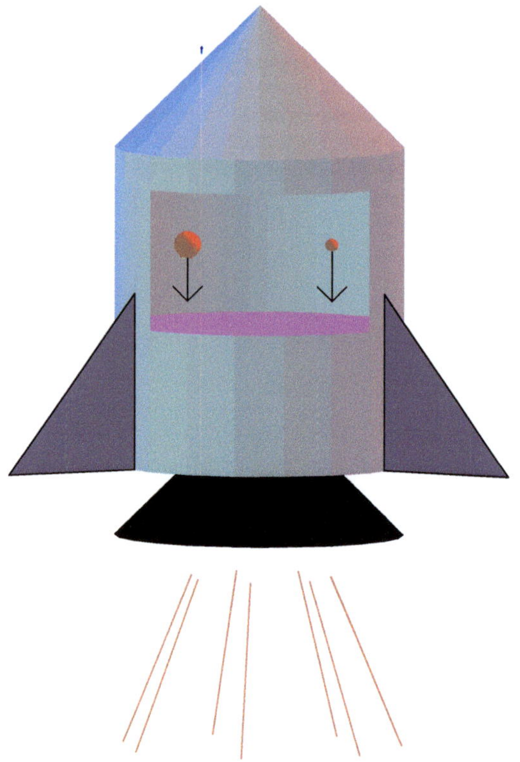

Fig. 4.2 A pebble and a rock are released from rest in an accelerating rocketship. Relative to the rocket ship as reference frame, they "fall" as if they had been released at the surface of the Earth

$9.8 m/s^2$. This equivalence is commonly encapsulated into a principle, the Principle of Equivalence which states that *a gravitational field is locally equivalent to an accelerated reference frame.* The gravitational field in this case is the field produced by the mass of the Earth and the accelerated reference frame is that of the accelerating rocket ship. Notice that the word "locally" has been slipped in to the statement of the Equivalence Principle. This is because the equivalence is only an approximation: In the first case, the pebble and the rock do not fall on precisely parallel lines but rather on converging lines that point to the center of the earth (Fig. 4.3). Thus, they land closer to each other than they were when they were released. On the other hand, in the rocket-ship experiment, they fall on strictly parallel lines.

Now let us consider another pair of Einstein thought experiments. Einstein is in that same elevator holding a rock in each hand at the same level and his wife M. Maric facing him, is holding two rocks, one above the level of Einstein's two horizontally-positioned rocks and one below this level so that the four rocks form a square. The elevator, at rest on the fourth floor, suddenly has its support cable severed and the elevator goes into free-fall. Just as it does so, all four rocks are released. What happens to the rocks? Ignoring for the moment the variation in the gravitational field of the Earth, all four rocks as well as Einstein, Maric and the elevator itself, fall

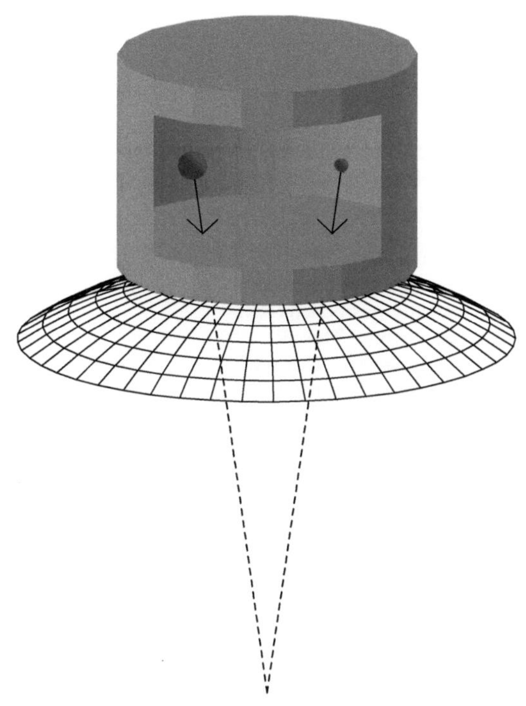

Fig. 4.3 Examining the motions of the pebble and the rock more precisely, we see that they fall on lines that converge towards the center of the Earth. Thus, they land closer together than they were when released

towards the ground at the same rate of acceleration, $9.8 m/s^2$.[3] So relative to the frame of the elevator, there is no relative motion among all of the aforementioned bodies; they remain together in a state of suspension.

Now we switch over to the elevator imbedded in the rocket ship. Einstein and Maric repeat the arrangement. While the rocket engines are shut, they release the rocks as before. Clearly by Newton's First Law, the result is as before in the freely-falling elevator: All the bodies remain suspended in their original positions. The Equivalence Principle has applied again. We have ignored the variation of the gravitational field so the effects of reference frame acceleration and the gravitational field have been realized: There has been a *local* equivalence.

Let us now look at the situation more precisely: When the cable is severed, Einstein's rocks, falling on lines that converge to the center of the Earth, come closer together with time. As for Maric's rocks, the lower rock, being closer to the center of the Earth than the higher rock, experiences a greater acceleration by Newton's law (5.1) (Fig. 4.4). (This effect manifests itself daily on Earth in creating the tides in the seas.) Thus, her rocks actually separate with the advance of time. The combination of both effects has been described by some as a kind of squeezing process.

We will return to the Equivalence Principle later but for now, let us advance towards the goal of formulating Einstein's theory of gravity, General Relativity. For

---

[3] We also ignore the very tiny interactions between the rocks themselves.

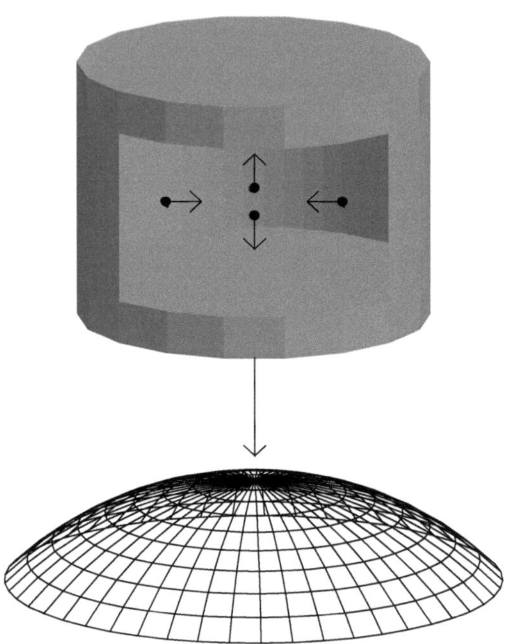

Fig. 4.4 Four rocks are released from rest within an elevator which is falling freely near the surface of the Earth. Examining the motion very accurately relative to observers who are also falling freely within the elevator, we see that while the horizontally-positioned rocks move closer together, the vertically-positioned rocks separate in time. This is because the lower rock experiences a slightly greater acceleration than the upper rock

this, we will require a little bit of mathematics. We return to the spacetime interval (3.27) of Special Relativity. When we moved with a constant velocity relative to the original coordinate reference frame, the spacetime interval retained its original structure, (3.28). However, it is easy to show that when we move with acceleration relative to the original reference frame, the form of the spacetime interval changes and depending upon the type of acceleration, often in a complicated manner.[4] Various combinations of the coordinates appear as multipliers of the dx^2, dy^2, etc. In fact, in general, new products of the form $dxdt, dydt, dzdt, dxdy$ etc. also appear. Let us write this as

$$ds^2 = g_{00}dt^2 + g_{11}dx^2 + g_{22}dy^2 + g_{33}dz^2 + g_{10}dxdt + g_{01}dtdx + .. \quad (4.4)$$

where we have used the symbol g_{ik} with i and k taking on the values $0, 1, 2, 3$ to represent these combinations of the spacetime coordinates that arise under the transformation. Here we are writing 0 for t, 1 for x, 2 for y and 3 for z.[5] Mathematicians and physicists express this in generality in the elegant compact form

$$ds^2 = g_{ik}dx^i dx^k. \quad (4.5)$$

---

[4] See for example [3] for an explicit demonstration of the changes to the form of the spacetime interval that are produced when the transformation to an accelerated reference frame is made.

[5] Note that for simplicity of expression, we have dropped the stars on the quantities that arise under the transformation to the new coordinate frame.

While our example has focused on Cartesian coordinates x, y, z, in General Relativity, we apply this form to arbitrary systems of coordinates.[6,7] The quantity g_{ik} is called the "metric tensor"[8] and it plays a vital role in Einstein's new theory of gravity. The reason is clear: The effect of acceleration is encoded in the functions that make up g_{ik}. But by the Equivalence Principle, gravity is locally equivalent to an accelerated reference system. Therefore, it is natural that this metric tensor plays a key role for Einstein's relativistic theory that includes gravity. Notice how it has evolved from Special Relativity which focused on the invariant nature of ds^2 in 3.27, so it has incorporated Special Relativity within itself properly as the limiting case. It is all very logical.

However, we must keep in mind that while we can effect a transformation from a situation with no gravity to a new accelerated frame in which there is the *semblance* of gravity, a second transformation that is the opposite of the first brings us back to the original situation of no gravity. You might be inclined to call this "hocus-pocus", and in a sense you would be justified. This is not "real" gravity; real gravity is the field that is produced by real bodies rather than by coordinate transformations. The interesting thing about real gravity, like the field around (and within) the planets, the Sun and every bit of matter and radiation, is that no coordinate transformation can remove it. In what follows, we will describe the essence of real gravity.

4.2 Towards a New Theory of Gravity

An important clue was already provided for us in our hunt for the new relativistic theory of gravity. This was contained in the fact that with real gravity, those rocks in the earlier example did not accelerate along parallel lines as they did for the simulated gravity, but rather when there was real gravity, they moved along converging lines. It leads us to ask: What is the difference between spacetimes that are the domain of simulated gravity as opposed to the spacetimes with real gravity? As we will discuss further, it will turn out that the spacetimes of simulated gravity are flat like the surface of a pane of glass whereas spacetimes with real gravity are curved, like the surface of a ball or the surface of a saddle. In fact General Relativity will bring us to the realization that while all bodies and fields (apart from the gravitational field) live *in*

---

[6] The repetition of an index means that it is to be summed over 0, 1, 2, 3. Here it is done for the indices i and k. This is referred to as the "Einstein summation convention". Note that the compact expression expands out to the form (4.4).

[7] Thus, x^i could represent (r, θ, ϕ) spherical polar coordinates, (r, z, ϕ) cylindrical polar coordinates, etc.

[8] For mathematically inclined readers, the vector, which is the mathematical object of greater familiarity, is actually a special case of a tensor. Like soldiers in the army, tensors are categorized by their rank. Rank number is given by the number of indices attached to the tensor. A vector, having the single index i, is a tensor of rank one. The metric tensor, having two indices, is a tensor of rank two. Tensors are mathematical constructs defined by their transformation properties. A nice development of the subject can be found in [7].

spacetime, gravity *is* spacetime, manifested by its curvature. Gravity is a "field" like none other.

For the benefit of the more mathematically-inclined reader, we provide in Appendix B a brief account of how the famous Einstein field equations of General Relativity are derived. These Einstein field equations

$$G^{ik} = \frac{8\pi G}{c^4} T^{ik} \qquad (4.6)$$

have been described by many physicists as the most beautiful and profound in all of physics. They connect the energies, momenta and stresses, the T^{ik}, of all matter and non-gravitational fields, to the metric tensor that is contained in the Einstein tensor G^{ik} in a very complicated way. And we recall that it is the metric tensor which describes gravity. Einstein might have wished for simplicity but ironically the Einstein tensor is anything but simple! A glance at Appendix B will reveal that it actually has 10 separate components, $G^{00}, G^{11}, G^{22}, G^{33}, G^{01}$... etc. with each one generating its own equation $G^{00} = \frac{8\pi G}{c^4} T^{00}, G^{11} = \frac{8\pi G}{c^4} T^{11}$... etc. These 10 equations are to be compared to the single gravitational field equation of Newtonian physics that we wrote in (4.3). The difference is still more dramatic: The Newtonian gravity equation has three simple terms on the left hand side whereas each of the 10 Einstein gravitational field equations has many terms with many that contain products of the components of the metric tensor g_{ik}.[9] Such equations are referred to as being "non-linear". When you read the expression "non-linear partial differential equation", you could substitute in your mind the expression "king-sized headache". This is because they are most often very difficult to solve. Generally speaking, exact solutions of useful non-linear differential equations have been found for the simplest of these equations. When confronted by more complicated equations of this type, approximation techniques are often employed to extract answers with hopefully accurate approximation to the exact solutions. Approximation procedures can be rather complicated and they have generated quite a bit of controversy over the years. While Einstein cherished simplicity, his visions embodying simple truths and observations translated into immensely complicated mathematical structures. That is the downside. However the upside is a wonderful new richness and a world of new possibilities that have excited researchers, and in turn the general public, to this day. In the succeeding chapters, we will explore various aspects spawned by Einstein's theory of gravity.

---

[9] A note for the calculus-equipped reader: These are actually partial derivatives of the metric tensor components.

4.3 Motion of Bodies in General Relativity

Let us return to Newtonian physics and (2.1) and (2.2). Taken together, they tell us that the acceleration a experienced by a mass m acted upon by the gravity of the Earth of mass M, depends only on M and the distance r of m to the center of the Earth. It's value is

$$a = GM/r^2. \tag{4.7}$$

From elementary calculus, we can express this as

$$a = -\frac{d}{dr}(GM/r). \tag{4.8}$$

We refer to GM/r as the "gravitational potential" of the Earth and the equation (4.8) says, in words, that the acceleration of a body under the influence of the Earth's gravity is given by the negative of the gradient of the gravitational potential. In Newtonian physics, the latter is the driver of gravitational acceleration. The "gradient", expressed by the symbol d/dr in (4.8), is just what the word implies: It is the steepness with which the quantity next to it is changing, here the gravitational potential, GM/r.

This is a particularly simple case. Generally, a distribution could well be more complex than the simple picture given by (4.8) in which the Earth is assumed to be a perfectly spherically symmetric distribution of mass. When the mass distribution is not spherically symmetric, the gravitational potential is accordingly more complicated. In general, we call the gravitational potential, ϕ, and we express the acceleration as

$$a = -\text{grad}(\phi) \tag{4.9}$$

where "grad" is the gradient. The simplest possible case would be one in which the potential is totally uniform, totally without change from point to point. In this case, the steepness with which it changes is zero, i.e. no steepness at all, the gradient of a constant is zero. If you walk along a level path, there is nothing to climb, no gradient whatsoever. This is equivalent to the situation in Newtonian physics where there is no force at all, where Newton's first law of motion comes into play and there is no acceleration,

$$a = 0. \tag{4.10}$$

Now with no force present and no gravity present, suppose we were to examine the motion from the vantage point of an accelerated reference frame. Then, relative to this new frame of reference, the equation of motion changes from (4.10) to what is called the "geodesic equation", where a new, general form of acceleration, a_{intrin}, the "intrinsic acceleration", arises and it is this new acceleration form that is zero for force-free motion. The new equation of motion is[10]

---

[10] See Appendix B for a more detailed description of this equation.

$$a_{\text{intrin}} = a + \Gamma uu = 0 \qquad\qquad (4.11)$$

where a here is the ordinary acceleration, Γ is the "Christoffel symbol" and u is the "four-velocity". The Γ is a geometrical quantity made up of the metric tensor g_{ik} and its derivatives. Of course when we apply the special case of using a non-accelerated reference frame in Cartesian coordinates, the Γ vanishes and (4.11) reverts to (4.10) as it must.

The Newtonian approach to motion is based upon the traditional idea of force, in this case gravitational force, inducing acceleration. The General Relativity approach is quite different. In General Relativity, motion under gravity alone is "free" motion, motion in which the new quantity, intrinsic acceleration, is zero. How does this arise? Actually, it is all very logical. After all, by the Equivalence Principle, the effect of gravity is locally like being in an accelerated reference system and mathematically, the motion in an accelerated reference system is given by the geodesic equation (4.11).[11] In fact the geodesic equation is the geometrical expression of the wonderful unifying principle in physics, the Principle of Least Action that we discussed briefly earlier. It is another case of an extremal equation. As a simple example, in three-dimensional geometry, the geodesic equation is the equation that guides one over the shortest path between two points. On the surface of a sphere, it picks out the paths along the great circles. For example, to fly from London to Vancouver, the pilots minimize the flight time (and in general, the expenditure of fuel) by following the shortest path, the route that cuts across the Arctic, tracing along the great circle. You can draw this path by choosing the plane through London, Vancouver and the center of the Earth and marking the plane's intersection with the Earth (Fig. 4.5).

Now the interesting connection to the weak gravity that we experience for example, on Earth, is this: The geometric quantity Γ approximates the gravitational potential gradient $grad(\phi)$. When it is simply shifted to the right hand side of (4.11), we get back our Newtonian equation of motion (4.9). It all fits together as it must.

There is a dramatic change in our conception of gravity that has taken place in the process just described. The Newtonian idea of the motion of a body under gravity as being the effect of action as a result of a force, the gravity force, has been totally replaced. In General Relativity, motion under gravity alone is force-free motion! Instead of the Newtonian force-driven motion, it is now seen to be *free* motion along these very special extremal paths in spacetime called geodesics.

J. L. Synge [6], a very brilliant mathematical physicist, was a very careful analyst of the essentials in General Relativity. He decried with gusto, physicists' fixation on the Equivalence Principle, emphasizing correctly that the essence of gravity in Einstein's Relativity is embodied in spacetime curvature. The supposed essential equivalence of gravity to the acceleration of one's coordinate system that one often hears about is, as Synge has emphasized, incorrect. Acceleration in a spacetime devoid of gravity produces pseudo-gravity, not real gravity. Just as acceleration produces the

---

[11] In Appendix B, we discuss the mathematics behind the intrinsic acceleration and how motion under gravity alone in General Relativity follows the geodesic equation.

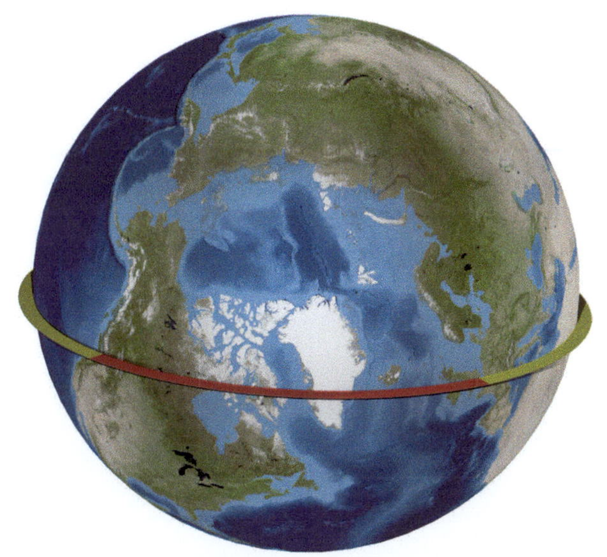

Fig. 4.5 The shortest path between Vancouver and London is the "geodesic" curve, the arc-segment of a great circle that is determined from the intersection of the plane containing Vancouver, London and the center of the Earth with the surface of the Earth. The Earth illustration is courtesy of NASA

semblance of gravity, the reverse of this acceleration removes it. Real gravity, i.e. spacetime curvature, cannot be removed by changing one's system of coordinates.

The distinction is essential and surprisingly, a fair number of researchers are either unaware or are unappreciative of this fact. Nevertheless, the usefulness of the Equivalence Principle as a *guide* should not be underestimated. It was this principle that directed Einstein towards the development of General Relativity. We have seen how it has led us to the metric tensor as the replacement vehicle for the Newtonian gravitational potential and how it has helped us to formulate the new General Relativity law for the motion of bodies under gravity. It has provided the scaffolding in the construction of the great edifice of General Relativity and once erected, there is no need to retain it as if it were an essential part of the completed structure. In this regard, Synge's point is well-taken.

In a variety of writings, you will witness authors proclaiming that once acceleration enters the picture, you are dealing with General Relativity, Einstein's theory of gravity. This is false. Bondi (see for example [2]) and others have produced excellent examples of Special Relativity formulations (i.e. Relativity *without* gravity) with accelerated reference frames. One must not equate acceleration to real gravity. Spacetime curvature is to be equated to real gravity.

Chapter 5
Testing Einstein's General Relativity

5.1 Motion of the Planets Around the Sun

It has been said that Newton was the first to state, with exemplary modesty, "If I have seen further it is only by standing on the shoulders of giants." While we would place Newton and Einstein, as the ultimate giants at the very highest level of achievement in the history of physics, there is much wisdom and justification in Newton's homage to earlier researchers. After all, the greatest advances were made with the help of important advances by those who preceded Newton and Einstein. Arguably the greatest of these was made by Copernicus who replaced the Earth with the Sun as the central body in what we now recognize as the Solar System.

For our purposes in leading towards the role of General Relativity, a useful starting point is the work of T. Brahe. He devoted much of his life to a careful study of the motions of the planets around the Sun. His precise measurements enabled J. Kepler to encapsulate the essential general truths that were buried in the vast stores of the Brahe data. These "Kepler Laws" are:

(a) The planets move in elliptical orbits around the Sun with the Sun itself at a focus of the ellipse.
(b) The planets traverse equal areas of their ellipses in equal amounts of time.
(c) The square of the period of the orbit of any given planet is proportional to the cube of the semi-major axis of its elliptical orbit.

Note that the elliptical orbit as indicated by the First Law has been exaggerated in the illustration Fig. 5.1. The planetary elliptic orbits are actually nearly circular. The orbits are more like Fig. 5.2. Far more pronounced orbital ellipticity occurs in the case of comets that are in bound orbits. However, by so drawing them, we are able to illustrate more dramatically, the nature of the Second Law, namely that in order to cover the same area at close approach as at the more distant orbital position, this law tells us that a planet moves faster at close approach than at the more distant position (Fig. 5.1). The Third Law tells us that the planets at greater distance from

F. I. Cooperstock and S. Tieu, *Einstein's Relativity*,
DOI: 10.1007/978-3-642-30385-2_5, © Springer-Verlag Berlin Heidelberg 2012

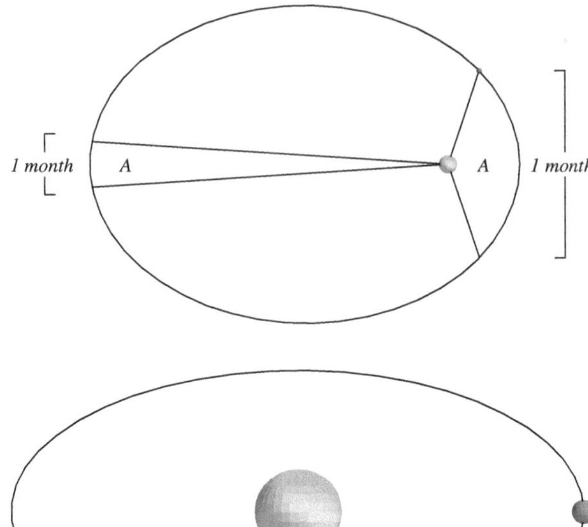

Fig. 5.1 To cover equal subtended areas in equal amounts of time, the planets must move more quickly at their closer approach to the Sun

Fig. 5.2 The actual planetary orbits are nearly circular

the Sun take longer to complete their orbits, that a Mercury "year", for example, is much shorter than a Neptune "year".

While Kepler condensed the vast stores of Brahe data into three empirical laws, Newton performed the ultimate further condensation. It came in the form of an actual theory of gravitation, that every body attracted every other body with a force proportional to the product of their masses and inversely proportional to the square of the distance between them. This is Newton's "Law of Universal Gravitation", as we saw in (2.2):

$$F = GmM/r^2 \tag{5.1}$$

with G, the constant of universal gravitation. It is quite remarkable to witness how Kepler's laws reveal themselves from Newton's great generalization. There are a number of excellent texts on classical mechanics to which the reader might wish to refer, that demonstrate this very clearly.

Newton's gravitational theory served physics very admirably for hundreds of years. It was Einstein's Relativity that created the necessity for its replacement. However even prior to Einstein, in spite of its enormous success as a theory, a lingering problem with Newton's gravity could not be overcome. It arose as a result of the detailed observations of the orbit of the planet Mercury, the innermost planet of the Solar System. The history is interesting. While Newton's gravity dictates that a single planet circling a star will trace out a *closed* elliptic orbit, the presence of additional circling planets in our Solar System must perturb the orbit of the planet

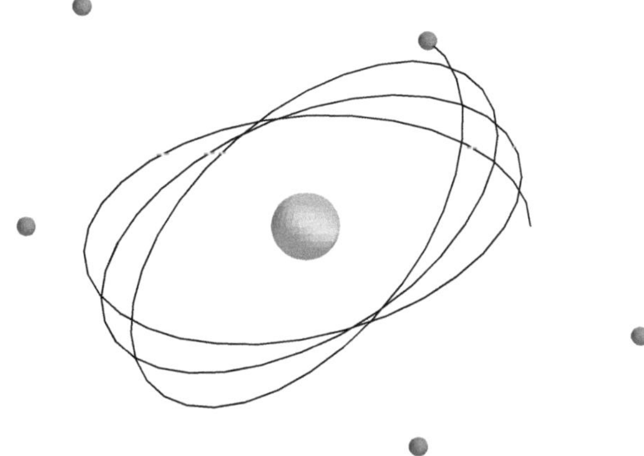

Fig. 5.3 The planets beyond Mercury's orbit cause the otherwise closed elliptic orbit expected on the basis of Newtonian gravity to precess as shown. General Relativity adds an extra amount of precession

in question. As a result of the extra tugs from the other planets, the elliptic orbit of any given planet will no longer be closed but rather will steadily shift or "precess" as illustrated in Fig. 5.3. By precise calculations, the orbital precession in the case of Mercury was supposed to be on the order of 500 s of arc per century. Surprisingly, astronomers found that the observed value was greater than the Newtonian-based calculated value by 43 s of arc per century. So convinced were astronomers of the infallibility of Newtonian gravitational theory that they posited the existence of a yet-unseen planet orbiting between Mercury and the Sun that would add an additional tug to resolve the discrepancy. It was even blessed with a name, Vulcan, However, try as they may, astronomers were unable to find any trace of Vulcan. Thus the matter was left in a state of abeyance. Later, we will revisit Mercury's anomalous precession. We will witness the magical power of Einstein's General Relativity to resolve the problem. To do so, we will focus on what is arguably the most important solution of the equations of General Relativity, that due to K. Schwarzschild.

5.2 The Schwarzschild Solution

Recall the old narrative of Newtonian physics: Bodies exert forces upon each other in accordance with the inverse-square-law (5.1) and by virtue of the gravitational force, they move in accordance with Newton's Second Law of Motion, $F = ma$. Einstein's General Relativity replaces this narrative with a wholly different vision: Matter curves spacetime according to the Einstein field equations and freely-moving bodies in the resultant curved spacetime follow the extremal paths of the curved spacetime, the geodesic paths. The first task is to determine the curved spacetime that the source under study creates. In the case of a star such as our Sun, it is a good approximation to consider it as a spherically symmetric distribution of matter. Accordingly, we choose the coordinate system best suited to take advantage of this

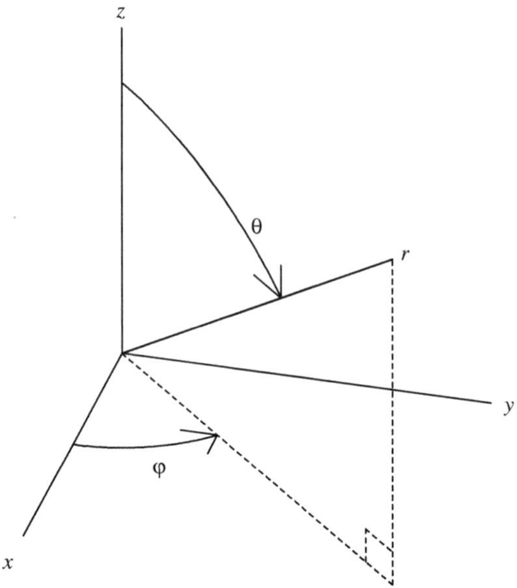

Fig. 5.4 Spherical polar coordinates r, θ, ϕ are the ideal coordinates to use for systems with spherical symmetry

symmetry, namely spherical polar coordinates (r, θ, ϕ). The origin of the coordinate system is taken at the center of symmetry, the center of the star, r is the distance from the center of the star and θ and ϕ are the polar and azimuthal angles respectively that determine the position of any given point. These are shown in Fig. 5.4.

Just as we were able to write the little increment of spacetime in terms of the Cartesian coordinate increments of physical distance as dx, dy, dz in (3.27), so too we can express it just as well in terms of spherical polar coordinates as

$$ds^2 = c^2dt^2 - dr^2 - r^2(d\theta^2 + \sin^2\theta d\phi^2). \tag{5.2}$$

There is an important idea that we have to get used to here. It concerns the translation from the mathematics to actual physical distance. When we express the little increments of distance in the x, y, z directions as dx, dy, dz respectively, we understand these as real physical distances. Now in spherical polar coordinates, the dr of (5.2) is the small increment in physical distance along the radial (r) direction. However the little increments $d\theta$ and $d\phi$ are *not* increments of physical distance. Rather, they are increments of angle, the small shifts in polar angle and azimuthal angle respectively. The small *physical* displacement in the θ direction is $rd\theta$. You will recall from your elementary geometry that arc length of a portion of circle circumference is equal to radius times the angle that is subtended and if you wish to encompass the entire circumference of a circle, you subtend the angle all the way around the circle, 360° or 2π radians of angle. Then all the $rd\theta$ increments add up to $2\pi r$, the circumference of the circle. Similarly, the physical increment of distance in the azimuthal ϕ direction is $r\sin\theta d\phi$. We must use the entire quantity, not just the little angle increment.

Fig. 5.5 In spherical polar coordinates, the small increments of distance in the radial (dr), polar ($rd\theta$) and azimuthal ($r\sin\theta d\phi$) directions are indicated. Multiplied together, they form a small volume element $dV = (dr)(rd\theta)(r\sin\theta d\phi)$

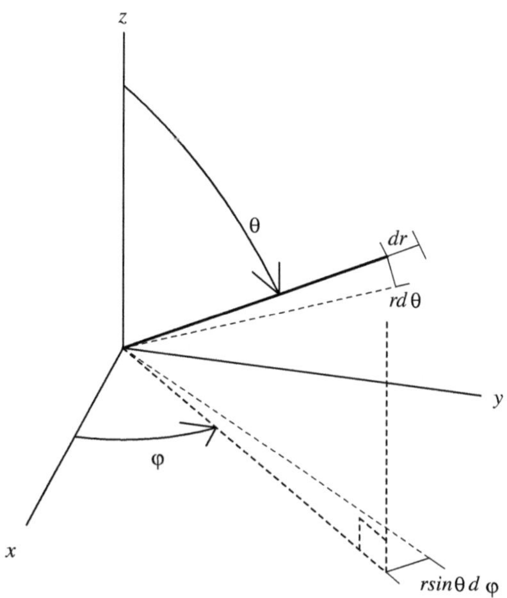

Of course the increment of distance traveled by light in the increment of time dt is cdt, the velocity times the time increment. We will be referring to this idea in an important manner in what follows for General Relativity.

We focus on determining the curved spacetime *outside* of the star where there is no matter, where the matter tensor $T^{ik} = 0$. The Einstein field equations (4.6) reduce to

$$G^{ik} = 0. \tag{5.3}$$

Within 2 months following the completion of General Relativity in late 1915, the solution of the Eq. (5.3) in the case of spherical symmetry was found exactly by K. Schwarzschild, and his solution had remained one of the key sources of further research in General Relativity over the subsequent years. As non-linear differential equations go, today we see these field equations for spherical symmetry as being very simple. This is a reminder that it is unfair to disparage advances made by others after the fact as being "obvious" or "trivial", adjectives often bandied about by some researchers who really wish that they had been the ones to have made the obvious and trivial advances! Schwarzschild is to be fully commended for seeing what others before him had failed to see. His name and work live on. Schwarzschild's famous metric is[1]

$$ds^2 = (1 - 2m/r)dt^2 - (1 - 2m/r)^{-1}dr^2 - r^2(d\theta^2 + \sin^2\theta d\phi^2) \tag{5.4}$$

---

[1] Readers with a more mathematical bent might wish to see the explicit form of the field equations for spherical symmetry and how they are solved in [3].

where we have adapted our coordinates to make $G = 1$ and $c = 1$. In the standard coordinates, where $2m/r$ appears in (5.4), we would have $2GM/c^2r$.[2] Note that if we set m to zero in (5.4), we recover (5.2), the spacetime interval of Special Relativity. It is most interesting to examine the effect of the simple introduction of mass on the spacetime interval. Recall that when we wished to express physical distance in the azimuthal direction, for example, we had to use the entire quantity that is multiplied by the azimuthal angle increment $d\phi$ in the spacetime interval, namely $r\sin\theta d\phi$. With gravity in General Relativity, it is even more involved, namely $\frac{r\sin\theta d\phi}{\sqrt{1-2m/r}}$. Similarly, in the radial (r) direction, the physical increment of distance is $\frac{dr}{\sqrt{1-2m/r}}$ and in the polar (θ) direction, it is $\frac{rd\theta}{\sqrt{1-2m/r}}$.

This brings us to what might seem a surprising next step but a step that necessarily follows: What was a small increment of time in Special Relativity, dt, is now, according to (5.4), $\sqrt{1-2m/r}dt$. We are led to the fact that relative to observers at different radial positions, periods of time actually change in the presence of gravity. From the perspective of an observer at large r, the clocks at an r value near to a star run slower than his clock and the increments of radial distance appear shorter. By contrast, from the perspective of an observer at small r value, clocks at large r value appear to run fast relative to his clock and radial increments of distance appear lengthened. In all of the cases discussed, we see that for a given r, the larger the mass m, the more dramatic the effect on the spacetime geometry, and the passage of time, with time being an element of spacetime, is no exception.

Recall the approach of General Relativity to motion: Matter and fields such as electromagnetic fields curve the spacetime according to the Einstein field equations. In vacuum, these equations are (5.3). Bodies in vacuum outside of these sources move in the curved spacetime according to the geodesic equations (4.11). For the curved spacetime in vacuum under the assumption of spherical symmetry, the metric is (5.4), the solution of (5.3).

When we substitute the solution (5.4) into the geodesic equations (4.11), we find, after some substitutions and simplifications, the equation of the orbit.[3] General Relativity provides an extra piece to the motion equation which is the lowest-order correction from Einstein's theory. If this term is neglected, the result is Newton's equation which yields the closed elliptic orbit for a single planet. The Einstein term provides the extra orbital precession.

---

[2] Setting $c = 1$ means that we are setting 3.10^8 m/s $= 1$, a pure number. This implies that 3.10^8 m $= 1$s. What this means is that with this choice, we replace every second that appears by 3.10^8 m in any subsequent point in the analysis. Similarly, setting $G = 1$ enables the replacement of every kilogram that appears by an appropriate number of meters. Thus, all quantities that follow are expressed in numbers of meters. While this procedure, highly favoured by relativists, might seem strange, it is really very sensible. Quite apart from simplicity of expression, it enables one to compare quantities meaningfully in terms of size as they all appear in terms of the same units, meters. In the end of the calculations, we can readily revert to conventional units if we wish.

[3] For the benefit of the more mathematically-inclined reader, the equation of motion is $\frac{d^2(1/r)}{d\phi^2} + 1/r = \frac{M}{l^2} + \frac{3m}{r^2}$ where l is the angular momentum per unit mass of the planet.

Over the years, we have witnessed the thrill felt by many students of General Relativity in seeing how the addition of the Einstein term in the equation of planetary motion almost magically provides that extra 43 s of arc per century of orbital precession for the planet Mercury. There are no free parameters involved in making the fit of theory to observation. Either it works or it does not work. And it works beautifully! It is hard not to become a firm believer in General Relativity when one witnesses this wonderful result.

It should be noted that a more dramatic demonstration of the precession is provided by the motion of the components of the binary pulsar $PSR\,1913 + 16$, a system consisting of a neutron star, the source of the observed very regular almost precisely periodically recurring pulses of radio-wave emission, in orbit with another very compact body, possibly a White Dwarf star. The orbit is very tight and hence the velocities are very high, yielding a more dramatic precession. Here the precession is in degrees per year rather than seconds of arc per century! Again, Einstein's General Relativity stands triumphant.[4]

5.3 Other Classical Tests of General Relativity

The correlation between the General Relativity prediction and the observation of the anomalous precession of Mercury's orbit is classified as one of the three "classical" tests of General Relativity. There are two other tests labeled classical. One is the gravitational red-shift, the difference in frequency of the light from emitting atoms in the Sun's photosphere as compared to their frequency as observed on Earth from like atoms. Earlier, we discussed the role of the observer's velocity in determining the period between the ticks of a clock in Special Relativity. As we saw above, in General Relativity, periods of time vary according to the local strength of the gravitational field: The stronger the field, the longer the period of time as judged by a distant observer. In other words, clocks run more slowly relative to a distant observer when they are in stronger gravitational fields in comparison to clocks that are in weaker gravitational fields. Since larger periods mean smaller frequencies, the light that is emitted from a given type of atom at the position of the Sun will have a lower frequency (and hence a longer wavelength) than that of light emitted by the same type of atom on Earth.

It is natural to ask how one would identify a given photon as having come from a particular atomic transition in a given type of atom. This is easy: When excited, each atomic species produces its own particular set of characteristic spectral lines, its own DNA, if you like. When you see a particular pattern, you might, for example exclaim, "Ah, it's sodium!" There could be no mistaking it for any other atomic emission, so particular are the "fingerprints" of any given atomic species. You know the pattern from observing the emissions from heating sodium in the lab here on Earth. Now one might think that by observing the emissions from the Sun, we would have an excellent

---

[4] We will have more to say about the binary pulsar when we discuss gravity waves.

means of testing General Relativity. Unfortunately, there are various complications. First, the Sun is a very agitated body whose emitting atoms are in continual turmoil. As a result, any shifts in frequency of the observed photons are due not only to the effect of General Relativistic clock-slowing but also to the Doppler-shifting of light coming from a moving source.

As an additional complication, there is a simple means of deducing the lowest-order gravitational effect without even invoking General Relativity [8]. It works as follows: Take a photon at the surface of the Sun with intrinsic energy $h\nu_s$ and effective potential energy due to its presence in the gravitational field and have it lifted to the Earth where it would have an intrinsic energy $h\nu_e$ and a new effective potential energy due to its location in a different level of gravitational potential. This is like throwing a ball up in the air: It starts with a certain speed that you give it in the throwing, a certain kinetic energy, and a certain energy because of the gravitational field that is present at the position at which you are throwing it. It rises in the air and as it does so, it slows down. It loses kinetic energy but gains gravitational potential energy, the energy that comes from being located at a higher point in the gravitational field. You can appreciate that it is a gain in potential energy because it has the potential to do more damage by dropping from higher up. So there is a trade-off between kinetic energy and potential energy but the essential thing is that the sum of the two forms of energy in each position is constant, *total energy is conserved*. This is one of the holy grails of physics. Now if you apply this kind of simple reasoning to a photon instead of a ball and you substitute its intrinsic energy $h\nu$ for the ball's kinetic energy (the photon always has the same speed c so its intrinsic energy shows up in terms of its frequency rather than its speed), and apply the principle of energy conservation, you get the lowest order effect that General Relativity predicts for the photon's frequency shift. Admittedly it is an unusual "mishmash" of physical ideas but it works to this level of accuracy. At this lowest order, there were very delicate experiments performed by Pound and Rebka (see [8] for references) to test this "gravitational red shift" effect. To bypass the uncertainties of source velocity of photons arriving from the Sun, they performed their experiments in a tower on the Earth where the source and absorber velocities could be carefully controlled and monitored. They relied upon the difference in gravitational potential at the base of the tower relative to the top of the tower. For their time, their experiments yielded a good level of accuracy in verifying the effect. Vessot and Levine were able to improve on the level of accuracy in a satellite experiment called Gravity Probe A.

The third "classical" test of General Relativity is the gravitational deflection of light. Again, thinking of a photon as a little ball of energy and therefore having an *effective* mass (not rest mass, a photon has zero rest mass) according to $E = mc^2$, when it passes by a massive body like the Sun, there will be a tug on it resulting in a perceptible shift in its direction of motion. Again, an amalgam of the quantum idea of a ball of energy with effective mass deduced from Special Relativity together with Newton's law of gravitational attraction would lead one to predict such an effect. However General Relativity, through its geodesic equation of motion (now applied to light) within the curved spacetime created by the massive body, predicts a greater degree of deflection compared to that predicted by Newtonian gravity.

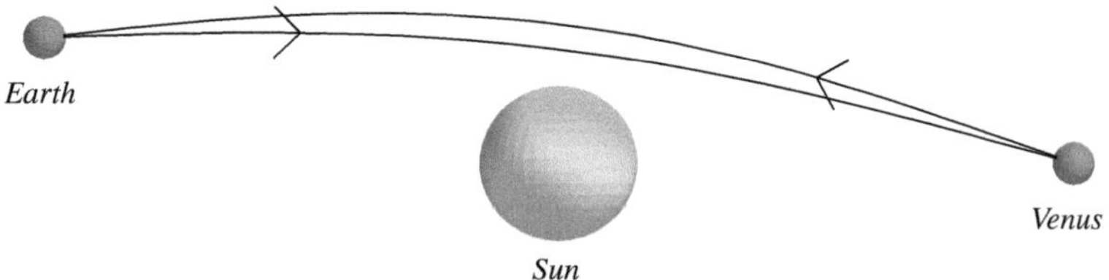

Fig. 5.6 Pulsed radar signals are directed to Venus during superior conjunction. The echo timing is used as a confirmation of the effect of General Relativity

In the early years after the advent of General Relativity, there was much interest in having this deflection measured. Would the observations confirm General Relativity or would Newton's gravity theory hold sway? However, there was a major obstacle for these observations. As they pass by the limb of the Sun, the very faint light from distant stars would not be visible to observers as the intense light from the Sun would dominate. (Try to see the stars in broad daylight!) However at a time of total solar eclipse with the Moon covering the bright disc of the Sun, the daylight briefly turns to near darkness and many stars are visible in the sky. By noting the positions of the stars in the given region, one could record their new positions during solar eclipse and so be able to note the resulting deflections.

Fortunately, during this period, a total solar eclipse was due to occur in Brazil and Africa. The very prominent astronomer, A. S. Eddington organized expeditions to the ideal viewing locations to record the deflected starlight positions. While there was some controversy surrounding the interpretations of the data (see, e.g. [5]), the results indicated in favour of General Relativity. At the time, there was much fanfare in the world's press regarding the results as a triumph for General Relativity. In more recent times, more accurate results have been claimed by observations of radio-wave deflections from quasars. These waves can be observed in daylight which greatly facilitates the observations. Moreover, with radio astronomy, there is the major benefit of much higher angular resolution, particularly with very long baseline interferometry using arrays of radio telescopes.

In the 1960s I. I. Shapiro initiated a new test of General Relativity, the "Fourth Test". This consisted of sending pulsed radar signals from the Earth to the inner planets Venus and Mercury during periods of "superior conjunction", when these planets are located on the side of the Sun which is opposite to the position of the Earth, as shown in Fig. 5.6.

The signals bounced off the inner planets and the time delay of the echo was recorded back on Earth. The delay predicted by General Relativity was consistent with the observed time delay. Even greater accuracy was achieved using interplanetary space probes with transponders receiving the signal from the ground station, amplifying it and re-transmitting back to Earth with a known phase lag via the directional antennae of the space probes.

5.4 Singularities and Black Holes

Let us return to the spherically symmetric spacetime in vacuum described by the metric (5.4). A spherical body like the Sun curves the spacetime around it and the extent of the curving is determined by the mass and compactness of the body, m: The bigger the mass for a given compactification, the greater the curving, i.e. the stronger the gravity. There is nothing stranger about this than what we experienced in Newtonian gravity which hinged, mathematically, on the Newtonian gravitational potential, m/r: The larger the m for a given r, the larger the potential. With this expression, we also see that the smaller the r, that is the closer we approach the body, the larger the potential. This told us that the Newtonian gravitational force, which derives from the gradient of the potential, grows with increasing mass and decreasing separation (Newton's inverse square law in a different guise).

What we had not discussed before is the issue of tracking r down to the value zero while maintaining m to be some non-zero value. With m being different from zero, we are not speaking about universal vacuum. Consider fixing m to be a very very tiny $1\,$kg spherical ball for example, and approaching it closer and closer. As we approach from $1\,$m to $1/2\,$m to $1/4\,$m to $1/8\,$m to $1/16\,$m to.. etc., the potential climbs from 1 to 2 to 4 to 8 to 16 to.. etc. When we finally let r approach to zero, we approach a number that we cannot even write down (even larger than a googol which is 1 followed by 100 zeros). We give it a name, "infinity" and we assign to it a symbol "∞". Of course our being able to contemplate this closer and closer approach assumes that we can shrink that ball to smaller and smaller radii. Generally speaking, when we have situations in physics such as with the potential where r goes to zero and the potential goes to infinity, we say that we have a "singularity". The standard view in such cases is that physics has figuratively hit a brick wall in understanding what is really going on, physically-speaking and we have to "go back to the drawing boards". A new physical theory is called for.

Now insofar as the Newtonian potential example is concerned, we could dodge the issue by saying that infinity looms only when we approach a presumed finite mass m having zero size and no finite mass can have zero size. However, the issues become much more involved when we deal with Einstein's General Relativity. There, rather than examining the Newtonian potential m/r, we actually must confront this very same quantity in a new context, that of the Schwarzschild spacetime expressed by the metric (5.4). We see m/r enter here in the g_{00} and g_{rr} components as $(1-2m/r)$ and $-1/(1-2m/r)$ respectively. The g_{00} component goes to minus infinity as r approaches zero so our former issue of dealing with infinity at zero value of r remains. However, in addition there is a totally new issue that arises. At the *non-zero* value $r = 2m$ ($r = 2Gm/c^2$ in terms of the conventional coordinates), the g_{00} is zero and the g_{rr} is minus infinity.

In the earlier days of General Relativity right up to as late as the 1960s, physicists referred to the spherical surface of radius $2m$ as the "Schwarzschild singularity". In a very early paper, Einstein had considered a spherically symmetric swarm of test particles. He showed that as the sphere contracted, the particles moved with increasing speed, finally approaching the speed of light as the sphere shrunk towards the size $r = 2m$. Thus, Einstein saw the $r = 2m$ surface as a barrier that could never be breached. Years later, Synge [9] showed that such test particles do not actually lie on the light cone as they cross $r = 2m$ heading inwards to smaller distances from the center. By not lying on the light cone, they are not violating the key tenet of Relativity that regular material particles can never attain the speed of light.

Through the later years, Eddington followed by G. Szekeres, D. Finkelstein and M. Kruskal, presented new systems of coordinates in which there was no apparent singular aspect at $r = 2m$ in terms of the new coordinates. This underlined what Synge had shown from a different perspective. However, a new aspect was introduced in the process. These new systems of coordinates showed the spacetime as being time-dependent, yet there is nothing really changing at all when we are at radial coordinate r bigger than $2m$. There is no harm in this and it is perfectly understandable. For example, in the system of coordinates displayed in [3] (see also [1]), the coordinates used are those that are actually *attached* to observers who are freely-falling. To help visualize such a "comoving" coordinate system, imagine, for example, a grid with an observer stationed for all time at every corner of the grid. Now imagine that this grid is being expanded uniformly. Each observer maintains his grid coordinates even though the distance between every pair of observers grows with time.

Thus, for a person who is falling freely inwards toward the attracting body, his newly acquired r coordinate, let us call it r^*, always takes on the same value. The fact that he is continually getting closer and closer to the body is displayed by his ever-shrinking *proper* distance to the body.[5] Thus, we can appreciate that the spacetime metric as expressed in terms of this new r^* radial coordinate, is necessarily time-dependent.

Now there are some very interesting, albeit somewhat bizarre phenomena that emerge when we study this in mathematical detail (see for example [1]). First, while we can switch from the r^* time-dependent representation back to the familiar r radial coordinate description of the spacetime for r values greater than $2m$ (and thereby assure ourselves that the spacetime is really intrinsically static), this is no longer the case when r has the value $2m$ or less. For those values, the spacetime becomes intrinsically dynamic! No coordinate system can ever be found for which this interior region is time-independent.

What does this entail physically? Let us build a scenario to appreciate it. Suppose a spherical star[6] has reached a point of contracting whereby its radius has shrunk to the value $2m$. In that case, the standard picture is that the star will enter a phase of complete "gravitational collapse". All elements of the star will move radially inward and nothing can stop this process. Nothing can emerge from the interior r less than $2m$ values to the exterior r values equal or greater than $2m$, not even light. A "black hole"

---

[5] This is a reminder to us of the importance of proper measures in Relativity.

[6] For simplicity, we consider here the case of a non-rotating body. Rotation injects additional complexity.

is said to have been formed, "black" because no light can emerge.[7] Now it is well to ask what happens when the star's elements reach the center. All would agree that the $r = 0$ point is singular. The detailed analysis tells us that these elements cannot simply stop there because to do so would entail the stopping of time itself! Some have suggested a rescue for the situation by imagining that this material re-emerges as a "white hole" in another universe. The white hole is the black hole with the direction of time being reversed, i.e. running the movie backwards. In other words, instead of continually and necessarily collapsing, the white hole is the situation of continually and necessarily expanding. Others have theorized with a different conclusion: Prior to reaching the very center of the star under gravitational collapse, when a sufficiently high density of concentration is reached, it is no longer legitimate to treat the material of the star as having its physical behaviour described by classical General Relativity. The argument is that a new yet-unknown theory of quantum gravity will eventually be found which will save the day, so to speak.

Let us consider the $r = 2m$ surface from another vantage point, that of the observer who is falling in towards this surface. When we examine the situation from his point of view, we can gauge his options at each stage. The detailed analysis reveals that the light cones at each point of his position on his path change their shape as he falls inwards. Recall that no matter what he does, continue to fall freely or turn on the retro-rockets of his space-ship to resist the fall, his spacetime trajectory must always stay within the light cone. He can never even coincide with a light trajectory which lies on the light cone, let alone proceed beyond the light cone. When he is outside the $r = 2m$ surface, as the Fig. 5.7 shows, he has options. He can continue to fall ever closer to $r = 2m$ or he can blast his way towards larger r values, i.e. he can escape the awaiting doom. There exists a wedge of opportunity between the light cone boundary that he cannot cross and the trajectories that head towards larger values of r. However should he dally and decide to act when he reaches $r = 2m$, he is too late. To even maintain his $r = 2m$ value, his spacetime trajectory is one of lying right on the light cone. However only light itself can do this, not our courageous explorer! So at that point, he must fall in, crossing through the $r = 2m$ surface and continuing towards ever-smaller r values. In fact after crossing the $r = 2m$ surface, even light itself must proceed to smaller r values. Therefore our hapless explorer cannot even beam outward to his family and friends back home that this is "goodbye" forever. The $r = 2m$ surface acts like a kind of horizon, like the one that we observe when we look out at the sea, where boats fade from view. Now since we are dealing here with events in spacetime, the $r = 2m$ surface has come to be known as the "event horizon".

It is actually somewhat confusing to speak of a value of r when it is smaller than $2m$ as being a "where" since r then becomes a time-like coordinate rather than a space-like coordinate. To make sense of it, it is best to follow the process from the point of view of the coordinate system attached to the falling observers. Because of the bizarre circumstances that ensue for $r = 2m$ and (inwards) beyond, Einstein and Rosen

---

[7] S. Hawking has theorized that quantum-mechanically, black holes will actually evaporate particles that emerge to cross out of the $r = 2m$ sphere but the process is very slow for stellar-sized bodies.

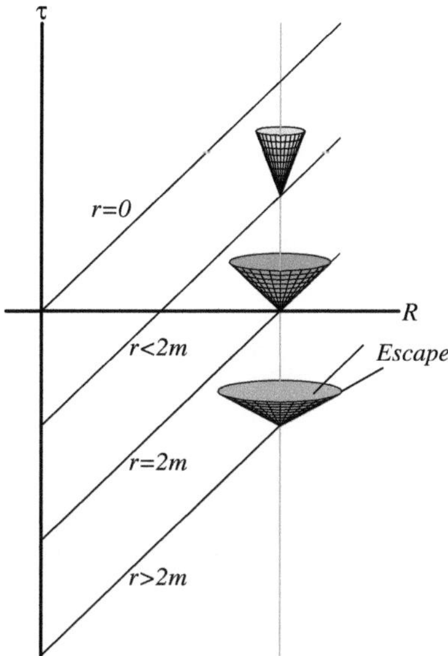

Fig. 5.7 The vertical line with constant value of R represents the path in spacetime of a freely-falling observer and the $r = constant$ 45° line represents the trajectory of an observer who maintains a fixed distance relative to the central body. *Light cones* are drawn at some points along the path of the freely-falling observer. The vertical axis is the time t axis and hence the history of an observer in spacetime is one that proceeds upwards in the diagram, i.e. with time advancing. Unlike the cases in Special Relativity, here using General Relativity, the light cones become thinner as r = 0 is approached. Before r = 2m is reached, an escape hatch is available. This is indicated in the figure as the zone between the $r = constant$ 45° line and the line tangent to the light cone. Motion in that zone is physically possible as it lies within the light cone and because it comprises directions with slopes exceeding 45°, it entails motion away from the body, achieving increasing r values. The observer could use rocket power to escape at that stage of his motion

developed a different kind of solution, one in which a copy of the Schwarzschild solution was joined to the existing one precisely at $r = 2m$. In this manner, the awkward region within $r = 2m$ was entirely excised out of existence. This join region or "throat" came to be known as the "Einstein-Rosen Bridge" (Fig. 5.8). These authors did not waver in their opposition to the idea that black holes are elements of physical reality, and their construction of the interior-free solution was an effort towards building a picture in tune with their conception of reality. In their view, $r = 2m$ was a physical singularity while the prevailing view is that it is merely a "coordinate singularity", an artifact of an inferior choice of coordinate system. In fact the prevailing view is that black holes are common occurrences of nature at the centers of galaxies as well as remnants of the collapse of very massive stars. If it should be the case that nature can present us with concentrations of mass for which further compression cannot be resisted and an event horizon can be reached, then the prevailing view has merit. However, it is conceivable that nature conspires to resist such formations. For example, we witness the phenomena of supernova explosions

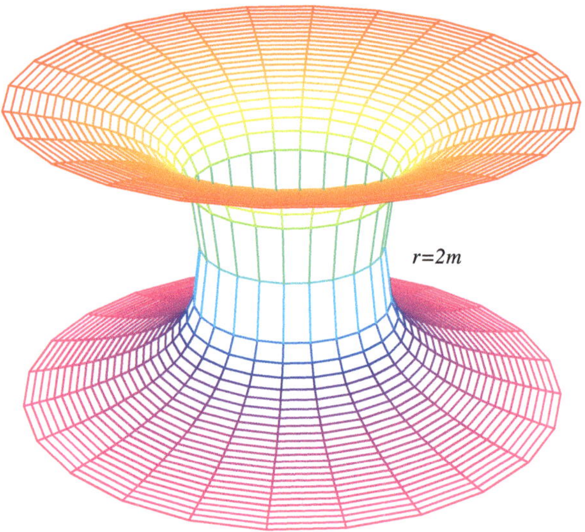

Fig. 5.8 The figure depicts the Einstein–Rosen bridge. Two Schwarzschild space-times are joined at the "throat", thus averting the troublesome region interior to the $r = 2m$ surface

wherein vast quantities of matter are blown away as certain massive stars collapse. Could these be instances of nature resisting complete gravitational collapse?

While it would appear that most researchers have become comfortable with both the theory and the reality of black holes, singularities are another matter. However, we discussed the problem of complete gravitational collapse which seems to indicate the inevitability of the eventual formation of an indisputable singularity, the kind related to having r go to zero while m remains non-zero in the instances where the expression m/r appears. So distasteful is the idea of such a singularity that the noted researcher, R. Penrose, devised an ingenious solution, now referred to as the "Cosmic Censorship Hypothesis". He conjectured that in the process of complete gravitational collapse, nature, acting as a censor, conspires to cast off any non-spherical elements as the body collapses so that what remains is an event horizon at $r = 2m$. The casting aside of such elements is necessary for his hypothesis, since it is only the spherically symmetric Schwarzschild solution among static vacuum spacetimes that allows the existence of a non-singular event horizon. For Penrose, the day is saved thereby since no information can be leaked to outside observers about the inevitable singularity that forms at $r = 0$. Outside observers are shielded from this embarrassment of a breakdown in physics.

However, there are problems with Penrose's solution to the problem. First, while outside physicists are spared the anguish of dealing with this now-hidden or "clothed" singularity, what about the physicist who falls in? Surely it would be wrong to dismiss him merely because he ultimately gets crushed to death. During the period that he is an observer, he is not shielded. Second, we have shown that even in spherical collapse, if the matter has an exotic equation of state of the form $P = k\rho$ where k is a constant lying between -1 and $-1/3$ (P is the pressure and ρ is the density), a "naked" singularity will form [10]. A naked singularity, as the name implies, is one that is not hidden from view. Outside observers can witness this singularity. While these types of equations of state for matter leading to the formation of naked singularities are not for the kind of matter that, as Bondi used to say, "you can buy in the shops", nevertheless they have an attraction all of their own. They are of the types that are believed to be behind the inflationary birth of the universe. Could there be some deeper connection? We are always on the hunt for clues that guide us to new insights.

Chapter 6
Gravitational Waves and Energy-Momentum

6.1 Introduction

In earlier chapters, we discussed sound waves, the actions of vibrating atoms and molecules passing along their motions in a linear chain which finally elicit vibrations in our ear drums. Our brains interpret these as sound, the information by which we communicate with each other and the pleasure that we gather from great vocalists and musical instruments that generate these waves in a skillful, blended manner. We also discussed electromagnetic waves, the transversely propagating electromagnetic fields which enter into every phase of our existence. At the most fundamental level, we understand their origin as deriving from the acceleration of a charge. Wiggle a charge back and forth and an electromagnetic wave is generated, flowing at speed c in vacuum away from the source. Depending on the frequency of oscillation, these waves create noise in our radios, warmed food in our microwave ovens, heat in our skins, vision in our eyes, etc.

As charge is to electromagnetism, mass is to gravitation. Thus it was natural for Einstein to conjecture the existence of gravitational waves in analogy with electromagnetic waves. While we will be discussing various similarities to which one can point, it is well to consider their overall essential difference: an electromagnetic wave is a distinct field disturbance propagating *in* spacetime. This field consists of oscillating electric and magnetic components which are perpendicular to the direction of the wave propagation direction. A gravitational wave is also an oscillating propagating disturbance but the disturbance is *of* spacetime itself! It is not an add-on to spacetime. For gravitational waves, it was the acceleration of a mass rather than the acceleration of a charge that would be the source of its production. Moreover, there was a more fundamental reason calling for the existence of gravitational waves. After all, whether in Newton's gravity or in Einstein's gravity, the form of the gravitational field is dependent upon the distribution of mass: changing this distribution, even by the simple act of raising one's hands in the air, must lead to changes in the gravitational field throughout the universe. Moreover, to be consistent with Special Relativity, this change must not be pasted onto the entire universe instantaneously,

F. I. Cooperstock and S. Tieu, *Einstein's Relativity*,
DOI: 10.1007/978-3-642-30385-2_6, © Springer-Verlag Berlin Heidelberg 2012

for to do so would violate the basic tenet of a finite velocity of propagation of interactions. This is where Newton and Einstein follow separate paths. It is at this juncture that for Einstein, the concept of a wave of spreading gravity arises, making changes to the distribution of matter realized in ever-increasing distances from the disturbance as time passes.

6.2 Gravitational Waves in Einstein's Theory

The beauty of Einstein's theory of gravity is manifest in the manner in which these fundamental ideas that we discussed above blend together so naturally. To set the stage, we return to Newton's gravity and its fundamental equation connecting source of gravity, mass density ρ to gravitational potential ϕ, namely (4.3). From this equation, Newtonian gravity assures that any change in the distribution of mass density, even the dropping of a pin, affects the gravitational potential *instantaneously* throughout the universe. This is the nature of Newton's gravity equation. It derives from a very simple theory. The mathematics dictates this behaviour which runs counter to the highly successful theory of Special Relativity.

By contrast to Newton, the much richer Einstein gravity theory replaces the Newtonian gravitational potential by the metric tensor g_{ik} of curved spacetime, and the mass density source of gravity by the full energy-momentum tensor T_{ik} of all of the matter, stresses and fields other than gravity which are present. The Einstein field equation that replaces the Newton field equation is (4.6). Under the right circumstances, when the gravitational field is weak, and a particular choice of coordinates is made, the Einstein field equation becomes very similar, in approximation, to the Newtonian equation (4.3). This is as it must be because we must always remember that Newtonian gravity theory works very well to a high degree of accuracy in many situations. In the Einstein case, the metric tensor component g_{00} replaces ϕ and there is a very important addition of a calculus time-rate-of-change operator. This operator is responsible for making the changes to the matter distribution relay the information to the field in the form of a propagating wave. The wave moves outwards from the source at speed c in vacuum, in various ways analogous to the case in electromagnetism. The idealized sources of gravity waves first discussed by Einstein, Eddington and others nearly a century ago were simple mass oscillators (masses affixed to the ends of a spring and oscillating in a line) and rotating rods.

At first glance, given the analogy with electromagnetism, and given that the observation of electromagnetic waves produced by analogous mechanisms with charge sources are the routines of our experiences, one might be inclined to believe that these gravity waves would likewise be readily observed. However, in the almost century since their first consideration, there has never been a direct detection of a gravitational wave produced by a source in any laboratory and their only indicated physical existence derives from the very tiny changes in the periods of some carefully observed binary pulsar systems such as the first ever observed, PSR1913+16. The generally-held view is that these period changes are directly connected to a loss of energy from the system, energy that is carried away by the gravitational waves that are being generated. We will have more to say about this topic in what follows.

Given that there is such a fundamental phenomenon in the gravity wave concept and given the paucity of evidence for its very existence, it is quite understandable that many papers and discussions have been devoted to gravity waves over the years. A small number of researchers have even questioned their very existence. For reasons discussed earlier, we feel the necessity for their existence. As to why they have never yet been observed in an experimental set-up, a convincing case is made on the basis of two essential differences from the electromagnetic wave detection case. First, the gravitational interaction is inherently very weak. It is mediated by the constant of universal gravitation, G, which is a very small number in conventional units. As an example of what this entails, it is of interest to note that the electrostatic force between two electrons is on the order 10^{45} times larger than their gravitational interaction. That is 1 followed by 45 zeros! It is worth contemplating the enormity of this number to help us appreciate the impact that this would naturally entail. Second, with regard to the type of wave of strongest possible order, in the electromagnetic case it is the "dipole" wave whereas in the gravitational case it is the "quadrupole" wave.[1] To elaborate on their distinction would take us beyond the scope of this book. Suffice it to say that the latter are much weaker than the former. Thus, we are prepared for a challenge in dealing with gravitational waves on an observational basis.

6.3 The Energy Issue

The concept of energy in General Relativity has been problematic from the earliest years. Energy and its conservation has been a vital anchor for physics in general. When there arose a situation in which energy was seen not to be conserved, a new form of energy would be identified as the missing link and energy conservation would be restored. In classical physics we have energy of heat, of sound, of compression, energy stored in the electromagnetic field, in the gravitational field, etc. In Special Relativity, we discussed the inherent energy in a mass m given by mc^2. We have become quite comfortable with energy as a concept, particularly when it is nicely localized, as it is within the location of the mass in Special Relativity.

However, it is somewhat less attractive as a concept when we have to deal with the energy in Newtonian gravity. For example when two bodies with masses m and M are moving under their mutual gravitational attraction, a conservation law for energy is extracted from the sum of two kinds of energies, the kinetic energies of the bodies, $\frac{1}{2}mv^2$ and $\frac{1}{2}MV^2$ where v, V are the velocities of the bodies and a potential energy—GmM/r attributed to gravity. While we can feel comfortable with the "where?" question for the kinetic energies ("they are right where the masses are"),

---

[1] The mathematically inclined reader can read about these "multipole" aspects in [3].

the potential energy is not where the bodies are located but rather it is spread out all over space. Repeated exposure to the idea tends to induce a certain level of comfort (after all, those Newtonian mechanics homework problems work out so nicely) but when we step back and contemplate the situation more critically, a sense of unease can set in quite readily. Can we find a more intuitively satisfying approach?

We have hypothesized that energy in General Relativity, *including* the contribution from gravity, is localized in the regions of the energy-momentum tensor T_{ik} [11]. Recall that the energy-momentum tensor is the mathematical construct that embodies all of the different forms of energy *except* the contribution from gravity itself. In other words, by the hypothesis, energy is where the "stuff" is and the stuff includes not only the matter that we are used to, like rocks and water but also non-gravitational fields such as the electromagnetic field. Our hypothesis: the contribution from gravity is there as well.

You might object to this hypothesis on the basis that electromagnetic fields are spread out just as are gravitational fields, but there is the essential difference that is worth repeating: in General Relativity, all matter and non-gravitational fields are contained *within* spacetime but gravity *is* spacetime, manifested by its curvature. All of the standard "stuff" is an add-on to spacetime but gravity is not an add-on to spacetime-it *is* spacetime. Recognizing the difference makes it easier to accept the idea of such a localization. Localizing all the energy where the energy-momentum tensor is non-zero requires some fundamental re-thinking along the lines that the energy aspect attributable to gravity could well play a different role. You might feel it natural to propose that the location of gravitational energy is where the spacetime curvature is present, but we will provide reasons as to why this is not really the correct answer. The choice of localization in the regions of non-vanishing T_{ik} calls upon us to face the perhaps uncomfortable idea of a gravitational wave propagating in vacuum while not carrying any energy with it. How did we arrive at the idea of such a localization?

1. In both electromagnetism and gravitation, we considered the simplest possible waves, plane waves, waves that have the same structure and magnitude over a plane. Plane electromagnetic waves are described by the energy-momentum tensor T_{ik} and as a result, their energy density is always positive, regardless of the coordinate system in which we choose to examine them. They convey a force when they impinge upon a wall ("radiation pressure"). They enter our eyes, causing an excitation in our brains and as a result, we see. Their reality as an energy-carrying physical construct is indisputable. Compare this to the simplest possible gravitational waves, plane gravitational waves. These waves have their energy described by what is referred to as the "energy-momentum pseudotensor" t_{ik}. As the name implies, a pseudotensor lacks the important tensor property of "covariance", that is having a well-defined structure in every system of coordinates and which, if non-zero in any given frame, is non-zero in every frame. By contrast, the pseudotensor t_{ik} of the gravitational field can be made to be zero at any spacetime point that one might choose by an appropriate choice of coordinate system. And with plane gravitational waves, it is even more dramatic:

for them, one can choose a coordinate system in which this pseudotensor is zero everywhere in spacetime! [12] This fact makes it easier to accept the localization hypothesis. After all, how can one be confident of an energy of gravity in vacuum if the construct that is traditionally used to measure it can be totally wiped out for plane gravity waves simply by making the correct choice of coordinate system?

2. We considered the work of W.B. Bonnor [13] who analyzed General Relativity solutions for dust clouds that collapse without spherical symmetry, solutions found by P. Szekeres. Bonnor was able to match these Szekeres solutions to the Schwarzschild metric asymptotically, implying that the region within has a constant mass given by the m in the Schwarzschild solution. However, by the traditional approach to energy in General Relativity, such a collapsing cloud, moving *without* spherical symmetry and thus having a changing mass quadrupole moment, must be pumping out energy into the vacuum beyond the dust cloud. Again, this is problematic for the standard picture of energy in General Relativity.

3. While many researchers have been convinced that gravity waves carry energy because of the observed period change of the binary pulsar, there is a more fundamental alternative explanation: over half a century ago, in a largely unknown but important work, A. Papapetrou [14] showed that the field equations of General Relativity do not allow for the existence of purely periodic solutions. On this basis, the period-changing binary pulsar is simply manifesting its conformity with the mathematical demands of Einstein's General Relativity rather than the preconceptions regarding energy. We can view this as another indication of Einstein's Relativity being the ultimate key to the cosmos.

4. By our localization hypothesis, being connected as to location with the tensorial matter energy-momentum T_{ik}, gives the contribution from gravity to energy a tensorial basis. We will have more to say in this regard in what follows.

It is well to step back and consider what objections could be raised to this localization hypothesis:

1. *The need for a graviton.*
   Quantum theorists, working at the sub-microscopic level, see the world in terms of quanta of energy and angular momentum, tiny discrete units associated with elementary particles and fields. At the level of dimensions 10^{-10} m and smaller, nature is no longer seen as a continuum but rather one of lumpiness, and what was certainty of position and velocity is replaced by a bizarre world of uncertainty and probabilities. The quantum theory is wonderfully successful in practice, particularly as applied to electromagnetism, even though some of its most prominent practitioners admit to being at a loss to fully comprehend its fundamental underpinnings. The quanta of the electromagnetic field, called "photons", have discrete angular momentum $\hbar = h/2\pi$, h being Planck's constant, and energy $h\nu$, ν being the frequency of the underlying electromagnetic wave. This wave is constructed from a vector potential A^i. Because their angular momenta are a

single unit of $\hbar$, they are referred to as "spin 1" particles.[2] While in some experiments, electromagnetic waves behave as one would expect for wave phenomena, in various other experiments, the photons that are the quanta of the underlying waves behave like little balls with energy as we would visualize from classical mechanics.

The tensor that underlies the gravitational field is the metric tensor g_{ik} and because it has two indices i and k, quantum theorists ascribe a quantum to the gravitational field of "spin 2", i.e. an angular momentum of $2\hbar$ and an energy $2h\nu$ where ν is the frequency of the underlying gravitational wave. This quantum has been given the name "graviton".

While this is a logical continuation from the quantum theorist's natural viewpoint, it neglects the essential distinguishing fact that while all of nature's particles and fields are inhabitants of spacetime, gravity *is* spacetime. Thus, we should not be led to necessarily assume that the gravitational field must be quantized, as most particle theorists and others would have it. In fact there is no experiment that shows such a necessity. This is in stark contrast to the situation in electromagnetism wherein, at the turn of the twentieth century, there were various fundamental problems that were in contradiction with classical (non-quantized) electromagnetism:

(a) *The photoelectric effect, the phenomenon of electrons ejected from a metal plate by electromagnetic radiation shone upon it.*
Regardless of the intensity of the incident radiation, no electrons were ever emitted unless a threshold frequency was used and even the weakest beam with the right frequency would suffice to eject electrons. Such behaviour is inconsistent with classical electromagnetism.

(b) *The "ultraviolet catastrophe".*
The spectrum of radiation intensity versus wavelength from a heated black-body, dropped precipitously at ultraviolet frequencies rather than increasing steadily as one would expect on the basis of classical electromagnetism.

(c) *The spectrum of quantized electromagnetic emissions from atoms.*
Discrete frequencies rather than a continuum of frequencies, were observed, again not as one would expect from classical electromagnetism.

From these phenomena, it was clear that classical electromagnetism was left wanting from the ultimate physical arbiter: the experiment. Physicists can invent the most beautiful of theories, theories built upon the most elegant of mathematics, but if nature does not agree with its predictions, the theories rather than the experiments must be discarded. Thus it came to pass that classical electrodynamics was confined as a theory for the beyond-atomic domain and had to be supplanted, ultimately with quantum electrodynamics, in the atomic and sub-atomic domains.

---

[2] What we ordinarily think of fundamentally as particles, such as electrons which are also quantized entities having "spin 1/2", i.e. intrinsic angular momentum $\hbar/2$, actually can also be regarded as waves. This wave-to-particle association is referred to as "wave-particle duality", an insight tracing back to L. de Broglie.

By contrast, when it comes to gravity, while some non-gravitational quantum phenomena can be sufficiently influenced by the gravitational field to show up in an experiment,[3] there is no experiment that tells us that *gravity itself* must be quantized. The latter would be a leap of faith driven by a quantum theorist's natural predilection rather than a logically-driven progression.

2. *The Feynman thought experiment*
R. P. Feynman devised a thought experiment to lend credence to the view that gravitational waves carry energy. He considered a pair of rings threaded through a bar, free to slide with friction and confronting an incoming gravitational wave train. Noting that free particles have their relative distance periodically changed if they are oriented perpendicularly to an incident gravity wave, Feynman reasoned that the ring separation would also vary periodically. Thus, with the rings sliding along the rod under friction as their interspacing varies, heat would be generated. After the wave train passes, the net effect would be a bar at a higher temperature than previously, confirming that a transfer of energy has ensued from the waves to the bar.

At first glance this certainly sounds like a convincing argument, but what has been overlooked is that the bar itself has been presumed to be unaffected by the gravity waves, that the bar, in effect, is a perfectly rigid body. Our earlier discussion laid such a notion to rest. Nothing escapes gravity because gravity is spacetime and everything inhabits spacetime, including the bar. Thus we must inquire as to how the bar is affected by the gravity waves. In spite of the bar not being perfectly rigid, (you could think of it as an extremely taut spring with a sufficiently strong pulling-apart, it will lengthen ever so slightly and it will spring back upon release) it is well to ask if there would be relative motion between the rings and the bar nevertheless. To test this, we replaced the bar by an idealized elastic medium modeled electromagnetically as follows: consider a section of the bar to be modeled as two parallel equal and oppositely charged capacitor plates with a high frequency electromagnetic wave bounced between them. The opposite charges provide Coulomb attraction while the bouncing waves provide electromagnetic radiation pressure. With the correct choices of charges and wave intensity, the forces balance and the system is in equilibrium. Changing the strengths of the balancing elements changes the degree of rigidity that is being modeled.

It is readily shown, moreover, that the system behaves like an oscillator. When the plates are pushed together and released, the increase in radiation pressure forces the plates back. They overshoot their equilibrium position and as the pressure diminishes, the Coulomb attraction brings the plates back towards each other [15]. In this manner, we have a simulation of an elastic medium that we can analyze with the Einstein equations in conjunction with the Maxwell equations

---

[3] This should not be seen as at all surprising: after all, quantum phenomena are aspects of the ingredients of our universe and they would most logically be regarded as being affected by their habitat, namely spacetime.

of electromagnetism. The result of the analysis reveals that the plates' separation follows the same distance variation as do two free masses at positions adjacent to the plates. The result is independent of the choice of balancing elements. Considering the free masses as the rings and the electromagnetic oscillator as the section of the rod, we conclude that there is no relative motion, i.e. no rubbing, hence no energy transfer.

6.4 Can Energy Be Localized?

Prior to the advent of General Relativity, the issue of whether or not energy can be localized did not receive much attention. While we discussed earlier the spread throughout space of the negative Newtonian gravitational potential energy, it was generally viewed as non-problematic, perhaps because its gradient gave the correct force exerted on a body within the framework of classical physics. Some interesting energy issues were discussed by Feynman [4] and were dealt with admirably in his characteristically engaging style, but no remaining issue outside of General Relativity raised any alarm bells. The most enduring source of difficulty and controversy arose in connection with energy-momentum in General Relativity.

Earlier we discussed the energy-momentum pseudotensor, the problem-plagued construct that had been used for gravity in General Relativity to parallel the successful energy-momentum tensor that had served and continues to serve for the rest of physics. Bondi [16] introduced an energy flux construct that he named the "News Function" as a supposed antidote to the pseudotensorial affliction that plagued General Relativity but Madore [17] and we independently [18] for a particular case, demonstrated the equivalence of the News Function to the energy-momentum pseudotensor. Thus, the Bondi construct did not provide the solution that was hoped for. Years later, Bondi [19] affirmed that energy in General Relativity must be localizable. In the years that followed, various authors arrived at a kind of intermediate stance, declaring that energy in General Relativity is "quasi-localizable". They developed rather complicated structures, structures plagued with problems of their own. One might be inclined to regard the notion of quasi-localization of energy as akin to quasi-pregnancy of a woman. Such is the nature of various solutions proposed at times by physicists in difficult situations to set their minds at ease.

We kept the faith in our localization hypothesis but what remained lacking was a tensorial construction that would serve us for General Relativity. Recently, M. Dupre [20] proposed that the Ricci tensor R_i^k was the required tensor. This was particularly attractive in that it is a tensor which we know, from the Einstein field equations, is different from zero only where the energy-momentum tensor T_i^k is different from zero, in accord with the localization hypothesis. In what follows, we will follow a sequence of steps that led us to what we regard as the final forms for energy-momentum in General Relativity [21]. It is instructive to do so in order to give you, the reader, the nature of the thinking that goes on in physics research. However, some more sophisticated mathematical concepts are involved and the reader who is unprepared for such is advised to proceed directly to Chap. 7.

We first ask if there is some aspect that would propel us to pursue the Dupre proposal. There is, in that it has an immediate appeal, knowing that Tolman [22] had shown many years ago that a stationary or static system with energy-momentum tensor T_i^k has total energy, including the contribution from gravity, of value

$$E = \int (T_0^0 - T_1^1 - T_2^2 - T_3^3)\sqrt{-g}dxdydz \qquad (6.1)$$

where $-g$ is the positive value of the determinant of the complete four-dimensional spacetime metric tensor (g itself being negative for spacetime). This is the standard form in which the Tolman formula is given, but by the Einstein field equations (4.6), this can be expressed in the simpler form

$$E = \frac{c^4}{4\pi G} \int R_0^0 \sqrt{-g}dxdydz. \qquad (6.2)$$

Thus, with the appearance of the Ricci R_0^0 component, we see that the Ricci tensor is a natural candidate for energy-momentum in General Relativity.

Now in (6.2), we have only the energy and we have only one component of a tensor, namely R_0^0. We seek tensor expressions and we want more than just energy; we also wish to incorporate momentum. The minimal way to achieve this and still maintain the Tolman energy expression for the special case of stationary or static systems is to create a new tensor-like expression

$$E_i^k = \frac{c^4}{4\pi G} \int R_i^k \sqrt{-g}dxdydz. \qquad (6.3)$$

From (6.3), we see that the component E_0^0 is precisely the Tolman expression for energy.

This is a good start but when we look further, we see an unusual mixture of elements. This concerns the aspects of volume. Earlier we spoke about proper measure of distance, the truly physical measure. When we extend this idea to three dimensions, we can express the truly physical element of spatial volume, the "proper volume" dV_p which can be written in terms of "coordinate volume" dV as

$$dV_p = \sqrt{3g}dV \qquad (6.4)$$

where $3g$ is the determinant of the 3-space part of the metric and where dV is the "coordinate volume", $dxdydz$ when we are using Cartesian coordinates, $r^2\sin\theta dr d\theta d\phi$ when we are using spherical polar coordinates, etc. In curved spacetime, dV is not the volume that is occupied physically, rather it is dV_p.

When we examine (6.3), we see the coordinate three-dimensional volume $dV = dxdydz$ but it is multiplied by $\sqrt{-g}$, i.e. the wrong metric determinant for three-dimensional space. This led us to extend the concept of energy-momentum in General Relativity, introducing a new construct, "spacetime energy-momentum" $E^{*k}_{\ i}$,

$$E^{*k}_{\ i} = \frac{c^4}{4\pi G} \int R^k_i \sqrt{-g}\, d^4 x \tag{6.5}$$

where $d^4 x$ is the four-dimensional spacetime coordinate-volume element. In Cartesian coordinates, $d^4 x = dxdydzd(ct)$, in spherical polar coordinates, $d^4 x = r^2 \sin\theta dr d\theta d\phi d(ct)$, etc. It was the mathematics that was guiding us to the new concept.

While at first glance it might appear to be a radical step, in fact it is rather benign when considered more closely. For a stationary system, the component $E^{*0}_{\ 0}$ is simply the Tolman energy multiplied by the time that it is being observed. However the new concept assumes a more interesting role for dynamic systems which we will discuss in what follows.

In $E^{*k}_{\ i}$, we have an integral of a four-tensor, R^k_i over proper four-volume which is as tensorial as one can get for an integral. Only tensorial elements are involved in its construction. While it reduces to the Tolman energy times the time of observation for stationary systems, it is well to ask whether it is merely the end result of a more complicated expression in the general case of a dynamic system. The answer is "no". For suppose we were to have a more complicated tensorial form in general with R^k_i replaced by, for example $R^k_i + S^{jk}_{ij} + \cdots$ where the dots indicate other second-rank add-on tensors. However, these add-on tensors must vanish for the particular cases of static or stationary systems in order to agree with Tolman's expression, and hence they must be of the form of a partial derivative in time, $\frac{\partial \cdots}{\partial t}$. While we can certainly construct forms of such S-type tensors (or variations thereof), in the form of covariant derivatives (to keep them tensorial) which will indeed have the required partial time-derivatives, they necessarily have in addition, partial space derivatives that do not vanish when particularized to non-dynamic systems. This contradicts the Tolman form and hence the Ricci tensor R^k_i is the unique possibility [21].

While (6.5) is as far as we can go with an integral incorporating covariant elements,[4] we can go one step further in expressing spacetime energy as a four-scalar. We do so as a logical progression from Special Relativity in expressing the energy of a body with four-momentum p^i. Suppose that body is viewed by an observer with four-velocity u^i. Then the energy observed can be expressed in (four-scalar) invariant form as

$$E = p^i u_i. \tag{6.6}$$

This E has the familiar value $mc^2/\sqrt{1 - v^2/c^2}$ where v is the relative speed between the observer and the body and m is its mass. Clearly, if the observer is chosen to be

---

[4] Being an integral of a tensor makes it non-transformable as a second rank tensor.

comoving with the body, $v = 0$, and then we have the Einstein iconic expression for E, namely $E = mc^2$.

Again, we follow the natural course for spacetime energy. We express it in the invariant form

$$E^* = \frac{c^4}{4\pi G} \int R_i^k u^i u_k \sqrt{-g} d^4 x \tag{6.7}$$

where u^i is the four-velocity of the observer as before only now it is with a field of observers. In [21], we have developed this further with the example of the gravitational collapse of a spherical ball of dust.

We have followed through with this new approach to energy-momentum in General Relativity, partially for interest in its own right, but also as an example for the reader of how we search for clues to advance research. The process is seldom straightforward. The path is generally laden with obstacles and curious twists but the reward, well-worth the effort, lies in the hoped-for advancement of understanding the nature of the universe and its workings.

Chapter 7
The Universe According to Relativity

7.1 Getting Acquainted with the Elements of the Cosmos

In the previous chapters, we have already witnessed some of the greatest marvels that Einstein's Relativity has provided for us. In this chapter, we will describe one marvel that many consider the greatest of all: the power of Relativity to describe the overall structure, evolution and ultimate fate of the entire universe. Can one imagine a more profound, weighty, intellectual achievement? To get us started, we will try to give the reader a real feeling for the basic elements and incredible distances that we deal with in cosmology.

We begin with the Earth, our home, the third planet from the Sun. It is a wondrous beautiful place that may harbour the only life in our Solar System, the system comprising the Sun plus the family of bodies that circle the Sun. Many of us have taken continent-size trips of approximately 4,000 km. Having a feel for this great distance, it is not a difficult stretch to contemplate the actual size of the Earth. A trip around the Earth is approximately 40,000 km in distance, about 10 times the continental-sized trip. Around the Earth we have the circling Moon. It lies at a distance of about 400,000 km from the Earth, about 10 times the already great distance for circum-navigating the Earth. Now let us contemplate that flaming hulk that is our Sun, the behemoth that keeps us alive with its tremendous radiance. It is so big that the Earth with our Moon in orbit about us could easily fit inside a sphere the size of the Sun!

To go beyond this, we use a technique that gives us a feeling for the next level of distances and sizes, the technique of scaling. Imagine that we were so big that the Sun would appear to us as a tiny burning pebble the size of the head of a pin. On that basis, how big would the Solar System appear to us? The answer might be surprising: at this pinhead-sized scale for the Sun, Pluto (now no longer regarded as a planet) at the outer edge of the Solar System would appear as a microscopic-sized speck 20 m from the pin-sized Sun. The other planets would appear as specks in-between with the larger planets barely visible to the naked eye. If you ever wondered why we were not undergoing collisions with the other planets, you can now appreciate why this is

F. I. Cooperstock and S. Tieu, *Einstein's Relativity*,
DOI: 10.1007/978-3-642-30385-2_7, © Springer-Verlag Berlin Heidelberg 2012

the case. There is a lot of space available in which those little specks can fly around safely without colliding with each other.

The usefulness of the pin-head scale continues. The Sun is a rather unexceptional star and the universe is teeming with stars. It is well to consider on the pin-head scale, where the closest star is situated relative to our Sun. The answer might astound you: the nearest pin-head sized star is about 50 km away! This gives you a picture of just how great a challenge it would be to visit a solar system nearby to that of our own. It also illustrates why we are not seeing stars smashing into each other as a common occurrence: there is a lot of room out there in which these stars can move!

Staying with the pin-head Sun scale, we next consider the galaxy which is the basic building block of the cosmos. Galaxies are conglomerations of literally billions of stars, assuming a variety of shapes, often beautifully symmetric disc-like shapes. At this scale, a typical galaxy would be about 300,000 km in diameter, roughly three-quarters of the distance from the Earth to the Moon! We have learned to picture this kind of distance, so we now have some feeling for the basic structure of a galaxy.

In order to go beyond the single galaxy, we must compress the scale even further for visualization. We imagine that the galaxy that we have just described is incredibly scaled down to the size of a dime, a coin roughly a centimeter across. A dime is actually a good visual aid since like a dime, a typical galaxy has a radius about ten times its thickness. We have frequently asked students to guess how far away the typical neighbour to this galaxy is located on this dime scale. Answers have ranged from 100 m to 400,000 km or more. They are invariably amazed to learn that on the dime scale, the neighbour is typically a mere meter away! The universe is teeming with galaxies in reasonable proximity, making it understandable that galaxy collisions are not very rare occurrences. The observations presented to us by astronomers of galaxies in collision are wondrous to behold.

Like the planets that avoid falling into the Sun by virtue of their circular motion about the Sun, the galaxies generally support their structure by being in rotation about an axis. With such an enormous structure, it should not be a surprise that a single complete rotation takes a tremendous amount of time, typically 100 million years. Galaxies tend to cluster in groups of widely varying numbers, from just a few to thousands. We describe the overall distribution of galaxies in the universe as fairly "isotropic": looking out into all the possible directions from our vantage-point, we see a more-or-less similar picture of averaged galaxy distribution. In what follows, we will discuss how this fits in with current ideas about the origin of the universe.

7.2 Evolving Views on the Universe

It is well to begin with some perspective. Throughout the centuries, humankind has entertained a variety of ideas about the nature of the universe that seem laughable to us today. However, the ideas then current were embraced with often fairly solid belief, even certitude. Very early-on, even some of the learned entertained the picture of a flat Earth with oceanic water dripping off the edges. Then came the true picture of the

Earth as a sphere, but the then-truism was one of a central positioning and a central importance of the Earth in the universe. The Sun and the stars were seen to revolve around us, and the stars were considered points of light of no great consequence, unlike the bright Sun. These ideas, that we now regard as exceptionally naïve, were the result of very limited information. As information has increased, knowledge has increased and many scientists regard the state of cosmological status-quo knowledge today as very near ultimate truth and wisdom. It is well to step back and remind ourselves that such was the attitude of many scientists at the various stages throughout the centuries.

This brings us to a very basic question concerning the universe: is it static or is it dynamic, evolving in time? The former view prevailed for many decades, in fact into the first three decades of the previous century. What is surprising to us is that even the great genius of Einstein embraced the static picture of the cosmos. We have always found this to be amazing. After all, gravity makes bodies attract each other and they will move unless they are held back. The basic building blocks, the galaxies of the cosmos, are not held back, at least not by anything that is at all apparent. Therefore it would seem eminently sensible to assume the view that the universe should be dynamic. Elementary, it would seem. Could it have been the desire for stability, the fear of the unknown catastrophes that one might have imagined in the face of an evolving universe that had so many scientists fixated on a static universe? Regardless, we are very aware of the tendency of many scientists to cling with fierce tenacity to prevailing wisdom. That being said, Einstein can be excused. His revolutions in other directions showed clearly that he was not one to fear new ideas. This is well-documented in [5].

Almost 200 years ago, H. W. Olbers presented a paradox regarding the universe. Even well before his time, the stars were recognized as suns and the universe was regarded as a sea of these suns stretching to infinity. A simple calculation shows that with this picture, the sky at night should be ablaze with light from the sea of starlight. However the night sky is dark. Hence the paradox arises. In more recent times E. R. Harrison and P. S. Wesson et al. showed that the darkness of the night sky is due almost completely to the relatively short age of the universe counting from the time of the Big Bang. This places a severe limit on the amount of light that the galaxies could have produced. While the expansion and the red-shift of the light also come into play for the dynamic universe, their effect, these authors have calculated, account for no more than a factor of two and hence would not affect the result qualitatively.

However, red-shift is a key factor for another reason: it provides the essential evidence that the universe is in a state of expansion. We discussed earlier how when two observers are moving away from each other, light received by one observer necessarily appears to be shifted in frequency towards the red end of the electromagnetic spectrum in comparison to the frequency as seen by the other observer at rest relative to the light emitter.

In 1929, E. Hubble announced that the light from the distant galaxies was red-shifted and that the degree of red-shift that he observed was proportional to the distance from us of the galaxy observed. With red-shift proportional to distance and also red-shift proportional to velocity, it follows that velocity is proportional to distance: the further a distant galaxy is from us, the faster it is moving away from us. This is codified into what is referred to as "Hubble's Law",

$$v = HD. \tag{7.1}$$

In (7.1), v is the velocity of a distant galaxy relative to us and D is our distance from that galaxy.[1] The H in (7.1) is called the "Hubble constant" although strictly speaking, it is not really constant as it evolves with the evolution of the universe. When people refer to the value of H, they are really speaking of its present-day value. It is more appropriate to call H the "Hubble parameter".[2]

It was the observations of galactic red-shifts that overturned the notion of a static universe. There was a widely distributed photograph of Einstein peering through a telescope with Hubble looking on, no doubt with a great feeling of joy. The Hubble observations convinced Einstein that he had been wrong about the static universe idea, and he described as useless the work that he had done to produce a static model universe. We will discuss this issue later.

We now turn to the problem of universe model-construction. To begin, we look back in scientific history. The earliest picture was one of the Earth being the center of the universe. Later when it was understood that the planets, including the Earth, circle around the Sun as center, the picture shifted to one in which the Sun was taken to be the center of the universe. Later, when it was appreciated that the myriad of stars in the sky were also suns, the idea of our position being one of dominance in the cosmos was completely overturned. If the ancient astronomers had observed the red-shifts of the distant galaxies (that they would have taken to be little pin-pricks of light rather than huge blazing conglomerations of suns) and importantly, that they were seen as such in all directions from our position as center, and had they understood the significance of red-shift, likely they would have used this knowledge to fortify their claim that we are the center of the universe! However from our present vantage point that our galaxy, the Milky Way, is in no way special, we now make the assumption that *on the average*, every observer in the universe should see the same averaged picture. This is codified into a principle, the "Cosmological Principle": "At a given time, the universe has the same averaged appearance for all observers". Since we see an essentially isotropic (the same in every direction) picture of the distant galaxies moving away from us with increasing speed with distance, so too should this be the case for all observers.

Recall in Chap. 4 how it is the geometry of spacetime that describes gravity in Einstein's General Relativity. As well, recall that it is the "metric" that determines the spacetime geometry. We require the metric for a spacetime that satisfies the Cosmological Principle, one that accords the same averaged picture of the universe for all observers at any given time. It can be shown that this is most easily expressed

---

[1] However it has been reported by H. Kragh that K. Wirtz and L. Silberstein had preceded Hubble by several years in discussing the possibility of a Hubble-like law.

[2] Because H is not really a constant, the law is not really a linear one. It is treated as a linear relationship as an approximation that is reasonably good for not-too-distant galaxies.

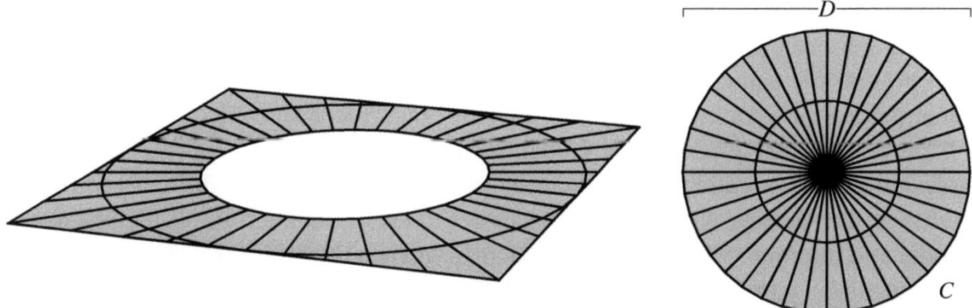

Fig. 7.1 This figure depicts the two-dimensional analogue of the spatially-flat $k = 0$ universe. The ratio of circumference to diameter of a *circle* drawn in this space has the value π

in spherical polar coordinates that we used in Chap. 5 in the form[3]

$$ds^2 = dt^2 - R(t)^2 \left[\frac{du^2 + u^2(d\theta^2 + \sin^2\theta d\phi^2)}{\left(1 + \frac{ku^2}{4}\right)^2} \right]. \qquad (7.2)$$

We need not be concerned about how this was developed nor the details about it. We have presented this equation to illustrate the power of General Relativity, how this rather simple metric can describe the geometry in averaged terms of the entire universe! The symbol k can assume the values 0, 1, or -1. For $k = 0$, the metric is nearly of the form of Special Relativity, i.e. no gravity at all, that we saw in Chap. 3. In fact when the R, which normally varies with time t^4 should be the special case where there is no change with time, R now being a constant can be folded in with u. We would then define Ru as the r that we had before and this gives precisely the metric of Special Relativity. So we see that for this special $k = 0$ case, were it not for the time evolution of the universe, with the R varying with t, we would have a very simple universe with no gravity. It is the time evolution that saves it from triviality! Many cosmologists believe that this evolving $k = 0$ case is what describes our universe. It is infinite in size, it has no boundaries and any snapshot in time gives the geometric appearance of flatness. Its geometry is the geometry of Euclid that we studied in grade school. If we were to draw a circle in the space, the ratio of the circumference to the diameter of the circle would measure π.

To get a better picture of this universe with time dependence, we resort to a two-dimensional analogue version. Consider a flat sheet of infinite size on which spots are painted equally spaced everywhere on the surface. We have homogeneity and isotropy- sameness at all points and in all directions. Imagine this sheet being heated

---

[3] Recall from Chap. 5 that ds is the tiny element of distance in spacetime, dt is the tiny increment of time as a coordinate, and $d\theta$ and $d\phi$ are the tiny increments of angle in the system of polar coordinates. Here du plays the role of the earlier dr.

[4] And so was written $R(t)$ to show the dependence.

by the same amount everywhere. As it expands, the spots are seen to increase their spacing between them. Clearly, any spot that we choose as center for observation will present the same picture: the spots around it all recede from the center in the same manner. The homogeneity and the isotropy are maintained.

The $k = -1$ and $k = +1$ cases are more interesting. The $k = -1$ case represents a universe of what is called negative curvature. The analogous two-dimensional model is that of the surface of a saddle. If you were to carve out a circle[5] at the center of a saddle and attempt to flatten it onto a plane, you would find that it would not do so without folding some excess material. Clearly, if you were to measure the ratio of circumference to diameter of this circle, you would find that it is greater than π, i.e. the circumference in this case is larger than it was for the corresponding $k = 0$ case. It turns out that this universe is also of infinite extent.

The analogue model for the $k = +1$ case is the two-dimensional surface of a balloon. Again, if you were to paint spots everywhere equidistant from their neighbour dots, you would see that for any point chosen as center, the same appearance is presented. The model for the universe here is particularly appealing: the expanding universe is represented by blowing up the balloon. As you do so, the spots separate but they do so uniformly. Any chosen center continues to experience the same view as any other chosen center. Again, the attempt to flatten the material cut out from a circle shape here will fail. Clearly, the material would have to be slit open to allow it to flatten onto a plane. There is less circumference for the given diameter of circle than there was in the $k = 0$ case. Now the ratio of circumference to diameter is less than π. We refer to this case as one of positive curvature. Again it is unbounded. Any journey on the surface never encounters an edge. However, unlike the other two cases, this universe is of finite size, the size of the surface of the sphere. Journeys in this universe can go on and on but one retreads the same territories, unlike the other two cases.

At this point, we ask the natural question: which of these three possibilities is our universe? As well, while we have the evidence that our universe is in a state of expansion at present (the observations of red-shift), we could imagine for any of the three possibilities that the universe could reverse direction at some point or even perform an unexpected "dance", reversing its direction followed by resuming its expansion, or even more complications in motion. The power of General Relativity is that it provides definite answers, answers which depend upon the averaged density of the matter in the universe. This is the point where the importance of the value of the Hubble parameter H enters:

(a) If the present averaged density of the matter in the universe is less than $3H^2/8\pi$, the universe would be of the negative spatial curvature $k = -1$ variety and according to the Friedmann solution of the famous Einstein equation of General Relativity, it would expand forever.

---

[5] The totality of points equidistant from the chosen center point is the formal definition of a circle. We can apply this here as we did in the flat space.

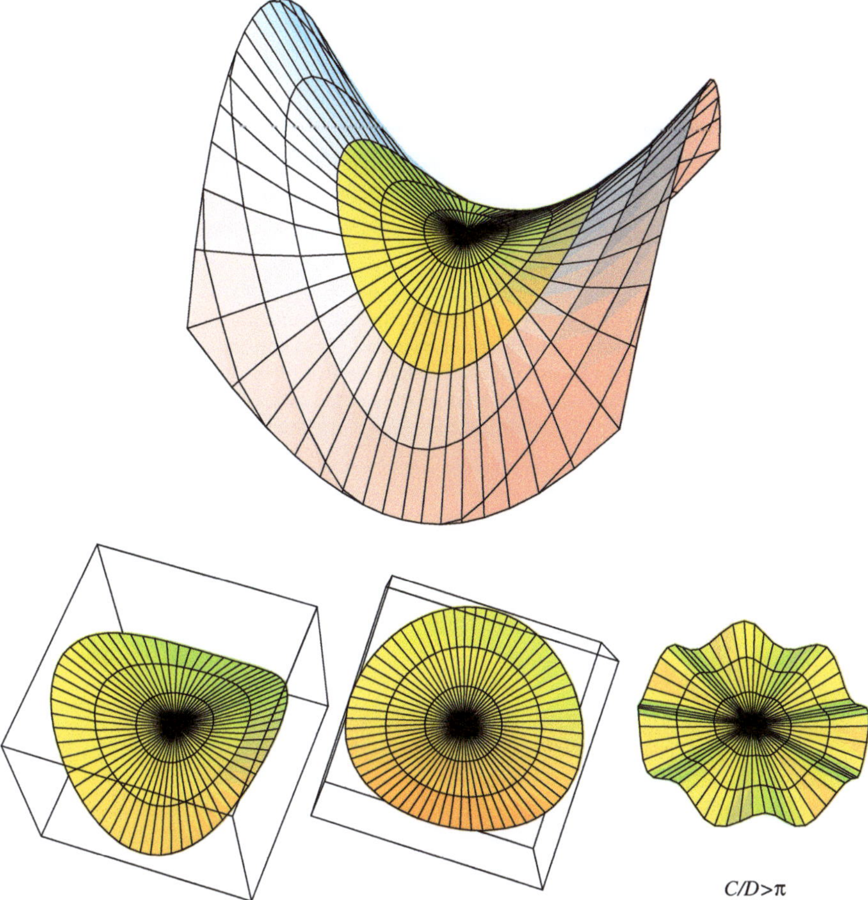

Fig. 7.2 In this two-dimensional analogue of the negative-curvature $k = -1$ universe, that of the surface of a saddle, the ratio of circumference to diameter of a *circle* drawn in this space is greater than π. The saddle-material has to be folded in segments in order to collapse onto a plane

(b) If the present averaged density of the matter in the universe is equal to $3H^2/8\pi$, the universe would be of the zero spatial curvature $k = 0$ variety and according to the solution of the Einstein equations of General Relativity, it would also expand forever. However in this case, it would approach a zero expansion rate as time approaches infinity. In both this case and the former case, the infinite universe would get ever less dense as time evolves.

(c) If the present averaged density of the matter in the universe is greater than $3H^2/8\pi$, the universe would be of the positive spatial curvature $k = +1$ variety and according to the solution of the Einstein equations of General Relativity, it would reach a maximum size at which point it would stop expanding and reverse direction. In this later contracting phase, the light that we would begin to

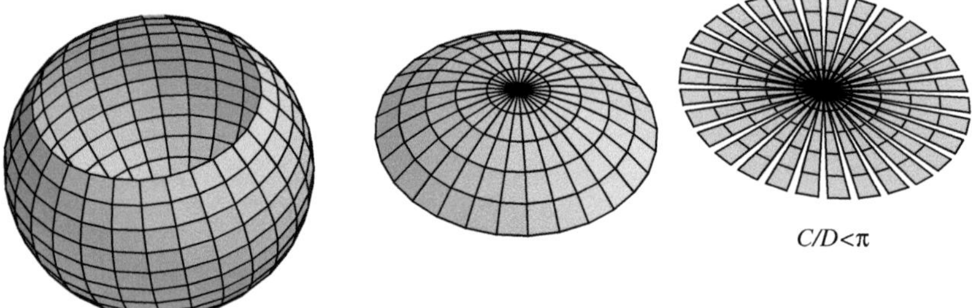

Fig. 7.3 In this two-dimensional analogue of the positive-curvature universe, that of the surface of a sphere, the ratio of circumference to diameter of a circle drawn in this space is less than π. For this model, in order to have the sphere material collapse onto a plane, it is necessary to make cuts in the material. The result is a sequence of wedges of material interspersed with no material

Fig. 7.4 The Einstein field equations determine that the *open* ($k = -1$) and the *critically open* ($k = 0$) universes expand forever whereas the *closed* ($k = +1$) universe will reach a maximum radius and then re-contract as shown

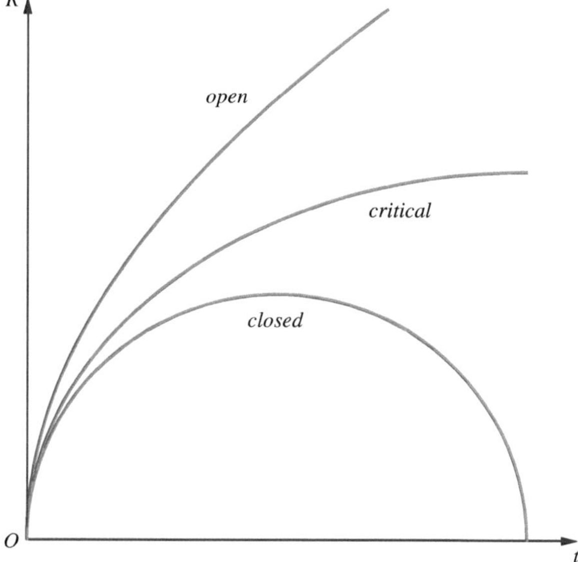

observe would be blue-shifted rather than red-shifted as it is now. And of course the nature of the universe would be so very different from the previous two cases. The $k = +1$ universe has a finite amount of matter and a finite size.

It is interesting to ask ourselves: which type of universe seems more "reasonable", an infinite universe with an infinite amount of matter or a finite universe with a finite amount of matter? Probably the majority of people would find it hard to entertain the idea that one could go on and on and on and find more and more and more matter, matter without end. They would more likely side with the more palatable idea that there is only so much matter out there and we would sample it multiple times in an ever-continuing journey of exploration. At the present time, the value of the Hubble parameter H would indicate a value of the density to favour case (a) as described on page 140. However, there is far more to this fascinating story that we will develop, as we proceed.

Chapter 8
Spacetime Diagrams for General Relativity

In the earlier chapters, we saw that much can be appreciated pictorially by suppressing two of the three spatial dimensions in order to bring time into the illustration for Special Relativity. We now do this with the universe itself where General Relativity comes into play. Gravity is the primary mover in cosmology, and this justifies our dependence upon General Relativity, our best theory of gravity, to describe the overall dynamics of the universe. We consider the origin O of our two-dimensional plot as the "big bang", the event of the birth of the universe[1] and instead of having the space coordinate as used previously to map out distance along x, one of the triad x, y, z of mutually perpendicular axes, we will use angles instead. The use of angles for plotting position is familiar to us: We describe our position on the surface of the earth using the angles of latitude and longitude. Where we sit, composing this book, our latitude is near to the 49th parallel, the boundary between the USA and Canada and our longitude is approximately $123°$.

For the evolution of the universe plotted in a spacetime diagram, we require one dimension to indicate time, so we have only one space coordinate at our disposal and for this, we use the angle coordinate on a planar slice of the universe. Rays fan out in all directions from the big bang O, from an arbitrarily set line as $0°$ all the way around to $360°$ where we return to $0°$. The matter of the universe bursts forth from the big bang along the rays as indicated in Fig. 8.1:

The twist here is that distance along the rays is measuring time since this is a spacetime diagram. In Fig. 8.1, we see the particles advance in time as we trace along the rays outwards, and here, the universe is in a state of expansion forever, the rays extending radially indefinitely. Cosmologists refer to this model as being "spatially flat". It is a universe that is infinite in size.

This spatially flat picture of the universe is one that is in favour with many, if not most, cosmologists at present. However, as we discussed previously, there are two other basic types of possible universe pictures that have been considered since the early dawn of General Relativity and here, we will focus on one of these, simulating

---

[1] We will have more to say about the universe in what follows.

F. I. Cooperstock and S. Tieu, *Einstein's Relativity*,
DOI: 10.1007/978-3-642-30385-2_8, © Springer-Verlag Berlin Heidelberg 2012

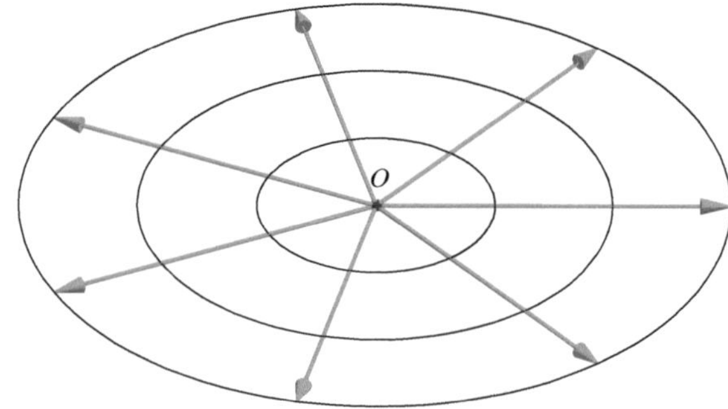

Fig. 8.1 The birth and everlasting life of the "spatially flat" universe expressed as a spacetime diagram

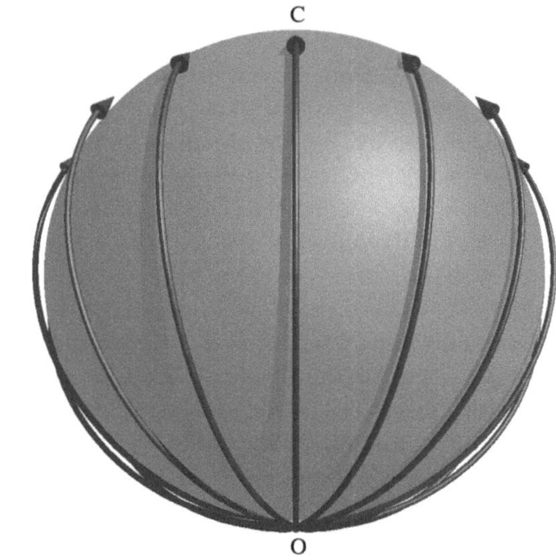

Fig. 8.2 This figure represents a spacetime model for a universe with simulated positive curvature. It is like a globe with longitude curves, O at the South Pole, C at the North Pole. *Arrow heads* pointing up along the curves indicate the advance in time

the finite "positive curvature" universe. Instead of a sheet, we now draw a spherical shell and place the big bang event O at the South Pole as in Fig. 8.2.

The curves of latitude that are drawn on globes of the Earth are now drawn with a different meaning: The matter bursts forth from O at the birth of the universe with the advancing latitude curve denoting the size of the universe that has been reached to that point. The universe elements spread out in all directions, advancing along the time lines. As they proceed in time, they pass the "equator" marker at which point contraction begins. The evolution continues as we get closer and closer to what we would ordinarily regard as the North Pole of the global Earth model. Finally, all of the elements come together at C, the "big crunch", which is the reverse of the big bang. Some refer to this as the death of the universe but others see this as a point at which rebirth occurs; a new big bang followed by cycles of expansion and contraction, continuing indefinitely. This raises new issues which we will discuss in the chapters ahead.

Chapter 9
The Motion of the Stars in the Galaxies

9.1 The Dark Matter Paradigm: An Overview

In the 1930s, F. Zwicky discussed the motions of the galaxies that comprised the elements of the Coma Cluster of galaxies. He noted that their velocities were not falling off with distance from the center of the cluster as would be expected on the basis of Newton's theory of gravity.[1] Years later, V. Rubin observed that within an individual galaxy, the velocities of the stars did not fall-off with distance from the galactic axis of rotation, again unexpected on the basis of Newtonian gravity. Rather, the stars essentially maintained the same velocity over the entire range from near the galaxy core to the visible galaxy edges. It should be noted that Rubin faced great opposition regarding her claims that the stars far from the galaxy centers were traveling at such high speeds. It was thanks to the emerging radio telescope technology in the 1970s that it became possible to measure with accuracy the velocities of the stars in the galaxies. These observations led to the idea that there was a great deal of matter that was not yet seen, which was providing the required extra gravitational action to make these motions possible, again according to the demands of Newton. Thus arose the theory that the universe contained a vast reservoir of "dark (exotic) matter", matter that did not have any electromagnetic interaction, unlike the normal "baryonic" luminous and non-luminous matter that constitutes the substance of our every-day world. This exotic dark matter could never make itself "visible", in the usual sense of the word. At present, there are observational research projects in progress in an attempt to find this elusive dark matter.[2] There are the observers at CERN in Geneva hoping to find new particles in their impressive ultra-high energy experiments. There are observations being carried out in very deep mine shafts to detect "WIMPS", weakly interacting massive particles, as potential candidates for dark matter. Thus-far, there has been no evidence for dark matter in these searches.

---

[1] We will return to this issue in the following chapter.

[2] In more recent years, the word "exotic" has been dropped in reference to the conjectured mysterious material and it is generally referred to simply as "dark matter". At times, we add the word "exotic" as a reminder.

F. I. Cooperstock and S. Tieu, *Einstein's Relativity*,
DOI: 10.1007/978-3-642-30385-2_9, © Springer-Verlag Berlin Heidelberg 2012

In fact, in the Gran Sasso Xenon100 experiment, a deeply submerged and very carefully isolated container of xenon has been observed for evidence of decays resulting from supposed WIMP-xenon nuclei collisions. The dark matter approach conjectures the existence of vast quantities of dark matter particles throughout the galaxy and given our motion in the galaxy, collisions with dark matter particles should be the routine occurrence, should these particles be present. The problem has always been one of distinguishing the dark matter effects from collisions with ordinary matter and it is the superb isolation afforded by the Gran Sasso mine that provides the required isolation. Recently it has been reported that the observations have reached a very high level of sensitivity yet no evidence for the existence of WIMPS, the leading dark matter candidate, has emerged.

Even prior to the yet-unsuccessful attempts to detect direct evidence for dark matter particles, a number of investigators found this dark matter prospect too akin to the ephemeral "ether" controversy of the previous century, a controversy ending with the discarding of its previously presumed necessity of existence. In place of the ether theory, there emerged the theory of Special Relativity and the ether vanished as easily as it entered the discourse of physics. Beginning with A. Finzi and followed by D. Milgrom, modifications of Newtonian gravity were proposed which were designed to explain the large stellar and galaxy velocities without the requirement for any dark matter. Milgrom's approach was put into a relativistic context by J. Bekenstein. J. Moffat and collaborators in turn produced new theory modifications in efforts to do away with dark matter.

We took an entirely different and, we would argue, the most conservative view of all, even more conservative than what many would regard today as the conservative view, namely that our conventional baryonic matter is dwarfed by vast amounts of exotic dark matter. Our reasoning had, at its base, the fact that General Relativity is universally considered the premier theory of gravity, and justifiably so. It had already demonstrated its stunning success in accounting for the subtle discrepancy in the orbital motion of the planet Mercury. Most importantly, the theory was not designed with the specific task of doing so; rather the number, the residual precession of Mercury's orbit, simply falls out from Einstein's theory without any "tweaking". For us, and probably for the many objective readers, this represents a very powerful endorsement, a sense that there is almost a kind of mystique attached to Einstein's great achievement.

The inevitable question arises: since Einstein's General Relativity has been appreciated for so long as our best theory of gravity, why was it not invoked before to analyze the motions of the stars in the galaxies and the galaxies in their clusters? The answer stems from the prevailing conviction that whenever the gravitational field is weak, and when the motions of the bodies involved are not approaching the speed of light, i.e. not "relativistic", General Relativity will hardly make a difference, that Newtonian gravity will provide the correct basic results. Since the motions in question are not relativistic and since their gravity is not very strong, the idea of applying General Relativity to these problems was never entertained previous to our work [23], at least not in print to our knowledge. In what follows, we will outline how we have applied General Relativity both to the issue of stellar velocities in galaxies as well as to galaxy velocities in clusters [24]. We will show that General Relativity presents many more possibilities than does the very simple Newtonian gravity, that its importance and richness in the galactic domain has been naïvely overlooked. We will show that General Relativity can indeed account for what is now known about the high velocities in the cosmos *without* vast quantities of exotic dark matter. Interestingly, it can also describe these motions with much more matter than is seen. This is not surprising because Einstein's gravity theory is far richer than that of Newton. We will develop this picture in what follows. We will devote a considerable amount of discussion to this subject as it brings together so much of fundamental importance to physics, astrophysics and cosmology.

9.2 Galactic Dynamics: Newtonian and General Relativistic Approach

Since the unexpectedly large velocities of stars in the galaxies have been the key factor in postulating the existence of dark matter, we will deal with this subject in some detail. Moreover, in keeping with the chosen targeted readership, insofar as reasonably possible, we will simplify any technicalities.

Recall from Newtonian gravity that the force between bodies of masses m and M separated by a distance r is

$$F = GmM/r^2. \tag{9.1}$$

Let m be the mass of a planet, treated as a test mass in the field of the very massive Sun whose mass is M. Then, invoking Newton's Second Law of dynamics, $F = ma$, in conjunction with (9.1), we have

$$ma = GmM/r^2. \tag{9.2}$$

For circular motion about the Sun, the acceleration a of the planet is related to the velocity and the distance as

$$a = v^2/r \tag{9.3}$$

where v is the orbital velocity. Canceling m on both sides of (9.2) and substituting v^2/r for a from (9.3), we have

$$v^2/r = GM/r^2 \tag{9.4}$$

which simplifies to

$$v = \sqrt{GM/r}. \tag{9.5}$$

So we see that according to Newtonian physics, the velocities of the planets fall off with distance from the Sun as $1/\sqrt{r}$. This would also apply to an extended spherical Sun with large radius. Over the years, astronomers applied essentially this kind of reasoning in the case of a galaxy to extrapolate that the stellar velocities should fall off as $1/\sqrt{r}$ as they viewed stars with increasing distance from the galaxy's axis of rotation, the axis which cuts symmetrically through the galaxy's central core. When Rubin announced her findings that the stellar velocities remained essentially constant as tracked outwards along the galaxy disk with increasing r, her results were regarded with great skepticism, so convinced was the scientific establishment of the expected $1/\sqrt{r}$ velocity fall-off with distance. As the evidence mounted, they could no longer ignore her work. For many years, her "flat galactic rotation curves", the plots of velocity as a function of radius r from the galactic rotation axis, have been well-accepted and frequently corroborated by quite a number of researchers.

As we noted above, in fact much earlier than the work of Rubin, Zwicky in the 1930s had come to a similar unexpected conclusion regarding the velocities of entire galaxies taken as individual unit constituents of the Coma Cluster of galaxies. He, as did Rubin, felt obliged to conjecture the existence of vast quantities of unseen matter that would act to drive these larger-than-expected velocities, matter now designated as "dark matter". The issues of the discovery of the particles that constitute dark matter and its essential nature have been regarded as issues of paramount importance in particle physics and astrophysics for quite a number of years and into the present day. This is understandable as the dark matter is generally believed to constitute approximately eight times the mass of the known conventional varieties of matter in the universe.

What surprised us was that researchers had accepted unquestionably the view that Newtonian gravitation was adequate for the analysis of both the motions of stars in the galaxies and of galaxies within galaxy clusters. They had been led to this stance because generally, for both types of phenomena, the motions are non-relativistic and the gravitational fields are weak. It has long been the accepted mantra that these conditions suffice for the legitimate application of Newtonian gravity theory. General Relativity, it was declared, could only present minor corrections. While in many physical situations, this is indeed the case as it was for the extra perihelion precession in the case of Mercury's orbit, there are reasons for the exercise of greater caution. For example, suppose we were to consider an ordinary rotating rod moving at non-relativistic speed or a normal linear mass oscillator, two masses affixed to the ends of a spring. If we were to accept the standard dictum, we would miss a most profound prediction of General Relativity that is totally absent in Newton's theory, the emission of gravitational waves. While the wave field is very weak at nearer distances in these cases in comparison with the dominant Newtonian component, the wave field gradient falls as $1/r$ with distance in comparison with the $1/r^2$ Newtonian component and hence the wave field ultimately dominates in the far-zone.

Another aspect concerns the non-linear structure of the Einstein equations and its role in sources whose dynamics are driven by gravity itself, rather than the much stronger forces that are actually ultimately electromagnetic in origin. Many years ago, Eddington [26] referred to this aspect when he developed the sequence of approximation equations for slowly moving weak-field sources. He demonstrated the importance of non-linearities when gravity itself is the agency responsible for the motion of the bodies.

These considerations led us to adopt an open mind when it came to the analysis of stellar motions in a galaxy. While there are a variety of phenomena at play within a galaxy, the essential dynamics are driven by gravity itself (what we frequently refer to as "free-fall", when gravity as the driver is dominant) and therefore we were particularly motivated to investigate whether there had been any appropriate theoretical development within General Relativity to model the dynamics of a system that could be seen as an idealization of a spiral galaxy. Fortunately, there had been indeed. Many years ago, van Stockum [27] developed the Einstein equations for a stationary axially symmetric system of rotating dust and Bonnor [28] had developed this work further. Their work was in the category of pure inquiry, the quest for probing General Relativistic solutions with different sources. For us, it was to recognize their idealized dust solutions as an excellent tool of transition to model the magnificent spiral galaxies presented to us in nature.

Now it is certainly true that the galaxies are not made of dust but rather of stars, planets, comets and a great variety of other debris. It is also true that the galaxies are not the idealized structures of perfect axial symmetry as studied in the van Stockum-Bonnor works. The galactic structures have a decided lumpiness with highly variable alterations of mass density, mostly in the form of stars, separated by near vacuum for the most part and they often are seen to have huge spiral arms, further negating the assumption of axial symmetry. However, General Relativity, as we have seen, presents a major mathematical challenge and hence while it would be most wonderful indeed to reach the level of modeling to fully reflect these aspects, we recognize that simplifications are necessary, at least at this stage in our development of the theory. The reality, as we understand it, is that the galactic constituents are primarily driven by gravity and the overall picture is one of a vast, roughly axially symmetric distribution of matter in steady rotation about the axis of symmetry, like the top that children pump into rotation and release. The primary gravity driver aspect justifies the modeling with dust, unlike the material of the child's top whose metallic elements are bound together by forces that are ultimately electromagnetic in origin. It is important to understand that given the nature of our modeling, there is no basis for expecting *detailed* concordance with astronomical observations. For example, some astronomers [29] have criticized our work on the basis that their observed density profile of matter at our location in the Milky Way is much greater than the density profile at the same radius in our model. We have offered a number of reasons for such a difference, quite apart from the essentially approximate nature of our model in [23]. First, the error bars in the data of [29] are very large and our solar neighbourhood could be a region having density considerably higher than average for our galactic radial position. Second, we have used a mere ten-parameter fit to model an average distribution. Third, without a knowledge of the densities at different regions of the Milky Way at the same r value, one cannot logically arrive at their conclusions. Fourth, quite apart from these factors, it must be remembered that the typical galaxy, albeit in basically gravity-driven motion, is a very lumpy distribution of stars and our model is a smooth dust *continuum*. In spite of all of these factors, in what follows we will present strong independent evidence for the essential concordance of our model with the reality of the galactic dynamics as they are observed in nature.

Astronomers would like to see the mathematical model present a galaxy more like the pancake-like appearance of the beautiful spiral galaxies that are observed. While further refinements of our dust model might very well get closer to this shape, it is quite possible that close matching is intrinsically impossible. The reason for this, as we suggested above, is that our model consists of dust rather than the lumpy distribution of reality. Thus, while our model captures the free-fall aspect, it fails to provide the additional localized enhanced tugs from each individual lump in the form of a star. Taking this lumpiness into account might very well bring the distribution into pancake-like distribution. Intuitively, one would imagine that dust would not tend to form pancake-like distributions but rather would tend to diffuse into more of an extended shape.

9.3 General Relativity Applied to the Observed Galactic Velocity Data

We will tread lightly on the mathematics of General Relativity to describe how the solutions are obtained for the galactic motion problem. First, the spacetime metric for our very idealized stationary axially-symmetric model is most naturally presented in the cylindrical polar coordinate system (t, r, ϕ, z) as shown in Fig. 9.1.

Having a coordinate (r) to measure distance from the galaxy's rotation axis and a coordinate (z) to measure distance above and below the galaxy's symmetry plane is very useful. Naturally as with the spherical polar coordinate system, we need the azimuthal angle (ϕ) as the third spatial coordinate. Note that there will be no ϕ-dependence in any of the metric functions because of the axial symmetry assumption. It is invariably helpful in choosing coordinate systems, as we have done here, to incorporate those coordinates that match the symmetries of the physical system being studied (Figs. 9.2, 9.3, 9.4).

It can be shown that the general form for a stationary axially-symmetric metric can be expressed as[3]

$$ds^2 = -e^{\nu - w}(udz^2 + dr^2) - r^2 e^{-w}d\phi^2 + e^w(cdt - Nd\phi)^2 \qquad (9.6)$$

where u, ν, w and N only depend upon r and z. It turns out that for weak fields, the u function can be chosen to be constant which we take to be 1 for maximum simplicity. The stationary aspect means no time-dependence and the axial-symmetry aspect means no ϕ dependence.

In problems of this form, there is an additional choice to make. We could have the coordinate system being non-rotating but it might perhaps appear surprising that it is most useful to have the coordinate system "comoving" with the matter, in this

---

[3] A fuller development of the mathematics that follows can be seen in [1, 23].

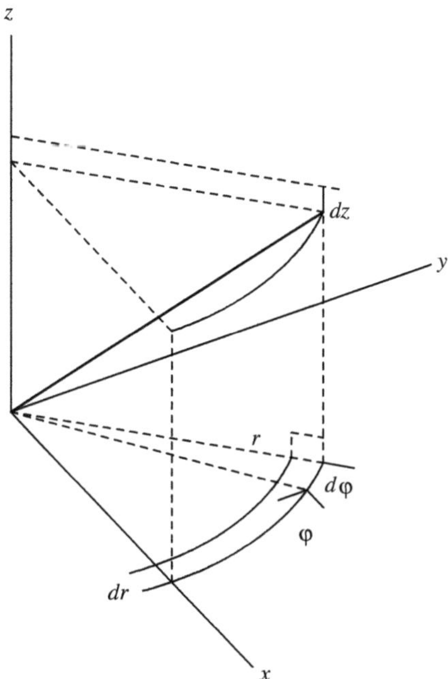

Fig. 9.1 A cylindrical polar coordinate system with coordinates (r, z, ϕ, t) is illustrated in the figure. The small increments of the spatial coordinates are indicated

case co-rotating. This choice, first made by van Stockum, actually simplifies the mathematics. Comoving coordinates are also used extensively in cosmology.

It is straightforward to show that the approximate angular velocity of the matter relative to non-rotating observers is simply related to the N function as

$$\omega = \frac{Nc}{r^2} \tag{9.7}$$

and the tangential velocity V of an element of the galaxy being studied is simply the angular velocity times the radius r at which the element is located,

$$V = \omega r = Nc/r. \tag{9.8}$$

After much mathematics, the essential Einstein field equations for the metric (9.6) simplify to

$$N_{rr} + N_{zz} - \frac{N_r}{r} = 0 \tag{9.9}$$

$$\frac{N_r^2 + N_z^2}{r^2} = \frac{8\pi G\rho}{c^2} \tag{9.10}$$

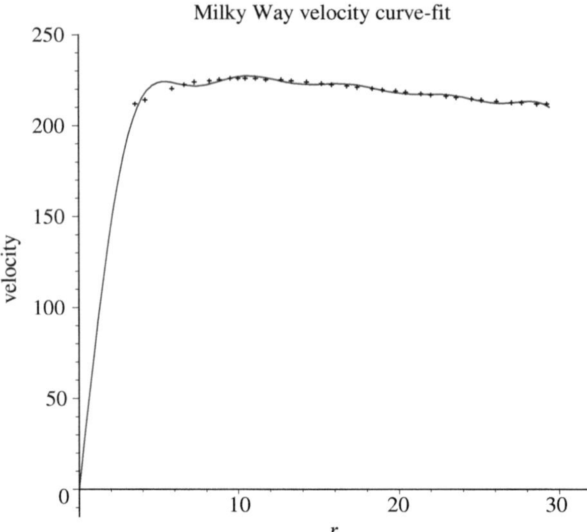

Fig. 9.2   Velocity curve-fit for the Milky Way in units of m/s versus Kpc

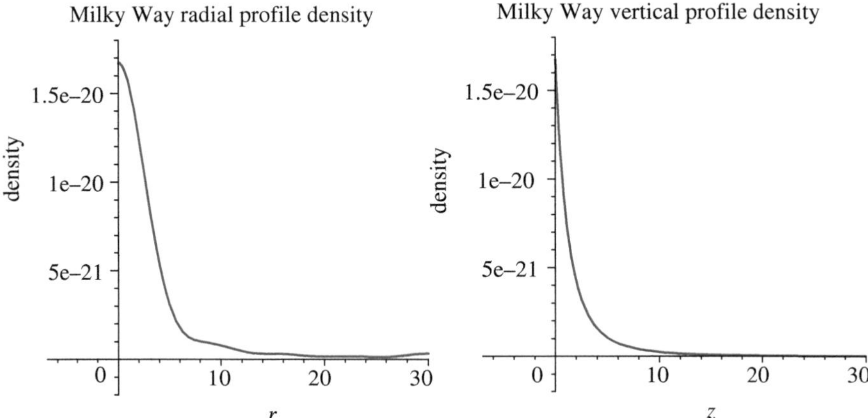

Fig. 9.3   Derived density profiles in units of kg/m^3 for the Milky Way at (**a**) $z = 0$ and (**b**) $r = 0.001$ Kpc

for the metric component N, and the ν function must satisfy the equations

$$2r\nu_r = N_z{}^2 - N_r{}^2, \quad r\nu_z = -N_r N_z. \tag{9.11}$$

In the compact manner that we prefer to write our equations as we have above, you might be lulled into thinking that these are quite simple. However, a subscript next to N, for example N_r means, for the calculus-trained reader, a "partial derivative"

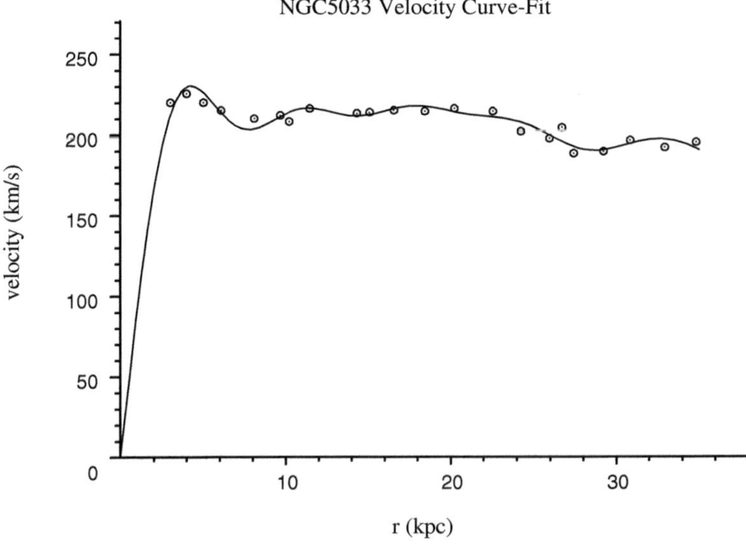

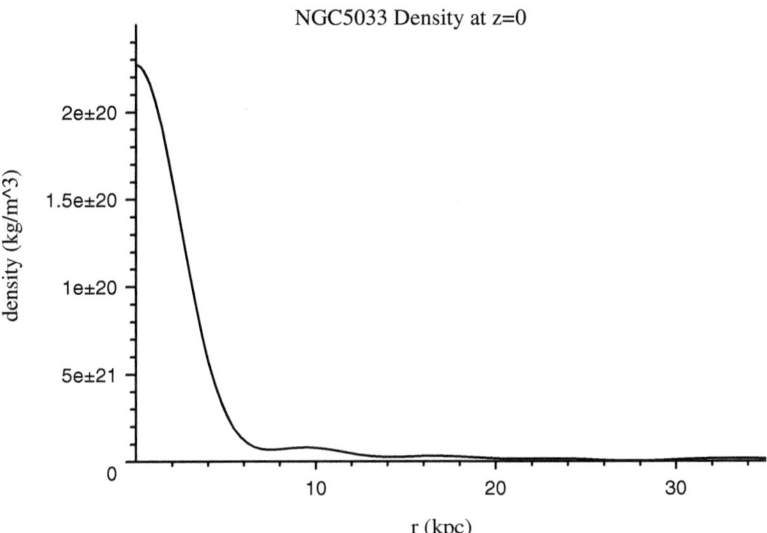

Fig. 9.4   Velocity curve-fit and derived density for NGC 5033

of N with respect to r, i.e. $\frac{\partial N}{\partial r}$ in the language of calculus and two subscripts mean two such operations. While this in itself is not generally a problem in "linear" equations such as (9.9), difficulties mount when such terms are squared as they are in (9.10) and (9.11). It is this *non-linear* aspect of the General Relativity field equations which make this theory so much more mathematically challenging as compared to the linear Newtonian theory of gravity. Quite apart from the fact that astronomers have invested so heavily in Newton's gravity, they are understandably reluctant to embrace the far more difficult Einsteinian gravity theory. However it is incumbent upon researchers to adopt the best that science can offer. Insofar as gravity is concerned, the preferred theory is General Relativity and when the preferred theory renders essentially different results, it must be chosen.

As it turns out, for the galactic problem, we are very fortunate that at least one of the Einstein equations, namely (9.9), happens to be linear. Linear equations are not only much simpler to solve than non-linear equations but also, they afford us the luxury of what is termed "linear superposition" of solutions. This means that if $S_1, S_2, S_3 \ldots$ etc. are solutions, then the linear superposition $C_1 S_1 + C_2 S_2 + C_3 S_3 + \cdots$ where the $C's$ are constants, is also a solution. It is this facility that we exploited in modeling Einstein solutions that conform to the observed galactic rotation curves.

For each studied galaxy, we chose the superposition of N function solutions of (9.9) to match the observed rotation curve velocity distribution according to (9.8). With N found, (9.10) determines the density distribution ρ as a function of r and z, i.e. throughout the galaxy. (From axial symmetry, it is the same for any azimuth angle ϕ value.)

In our first applications, we studied our own galaxy, the Milky Way, and three external galaxies NGC 3031, NGC 3198 and NGC 7331. These are discussed in detail in [23]. Because it is so "close to home", we present here the results from the Milky Way solution, that of our own galactic residence. More recently [30], we have analyzed three additional galaxies, NGC 2841, NGC 2903 and NGC 5033. We include here the table of results for the new set of galaxies as well as the plots for NGC 5033.

In the table of galactic data, the "critical radius" refers to the distance from the galactic center at which the last visible light is observed. Data beyond that radius is referred to as "HI" radiation. This data is in the radio-wave zone of the electro-magnetic spectrum. The "critical density" refers to the density that is read from the density plot at the critical radius. The "mass" values are those deduced from our General Relativistic model and "Kent's value" refers to the Newtonian-gravity-derived values listed in the Kent paper [31].

The first thing to note in our velocity curve fits is how well they agree with the observation data points. The next thing to note is the steep fall-off in density as we examine points further and further from the center, as we would expect for a good galactic model. It should be noted that these plots were realized with a mere ten parameters in the expansion of the solution, the $C_1, C_2, C_3, \ldots, C_{10}$ coefficients, as we discussed above. As well, we note that the masses that we compute are all smaller than those which have been deduced on the basis of Newtonian gravity. While we have listed here the table of results for the most recent study, [30], the same situation held as well for our previous study of galaxies [23].

Of particular interest is this: in our earlier analysis [23], we found that in the galaxies studied, the critical density was very near to $10^{-21.75}$kg/m^3 and we see this result again for the three new galaxies analyzed (see Table 9.1). *This concordance demonstrates the power of General Relativity to take from the data of observed stellar*

Table 9.1 Galactic data

Galaxy	Critical radius (Kpc)	Critical density (kg/m^3)	Mass ($M_\odot \cdot 10^{10}$)	Kent's value($M_\odot \cdot 10^{10}$)
NGC 2841	17.0	$10^{-21.60}$	51.5	74.6
NGC 2903	12.1	$10^{-21.59}$	8.9	19.4
NGC 5033	23.2	$10^{-21.83}$	22.2	37.1

velocity distribution in the galaxies to extract a connection with the essence of the galactic structure, a power that Newtonian gravity cannot provide. Moreover, it is to be noted that this agreement occurs for galaxies greatly different in their masses and at greatly different critical radii. Given the idealization of our model and given the small number of fitting parameters used, our results are as well-matched to the actual galaxies studied as one could possibly expect.

One can only marvel at the power of General Relativity to exhibit this concordance. Moreover it has reached the level of a prediction: *present a galaxy with a given rotation curve and our theoretical structure will predict the approximate radius of its optical edge.* Prediction followed by experimental/observational confirmation is what we seek in science to build confidence in the correctness of our theoretical efforts.

Chapter 10
The Motion of Galaxies in Galaxy Clusters

10.1 Preliminary Notes

Zwicky's observation of the unexpectedly high velocities of the individual galaxies in the cluster of galaxies known as the Coma Cluster led to the first appearance of the dark matter hypothesis. Unlike the organized motions in a nearly steady-state of the individual stars in a single spiral galaxy, the galaxies within a cluster are generally observed to move chaotically. Since we are not yet in a position to model such intrinsically time-dependent chaotic systems in General Relativity, we took the first step in the direction of dealing with time-dependence, again with an idealization. In what follows, we will describe our idealized model that brings into play intrinsic time-dependence. As a bonus, we are able to counter some of the criticism that had been levied against our earlier work.

First, unlike our idealized study of the stationary motion of stars in a single galaxy,[1] we are able to observe General Relativity's power to deal with systems that actually evolve in time. Second, unlike the solutions to the approximate Einstein equations that we employed previously, we are able to work with exact solutions. The latter is particularly gratifying as there has been a plethora of criticism over the years against a variety of research studies regarding the use of solutions to the approximate equations. Working with exact solutions removes all doubt so we were particularly gratified to be able to make the transition to exactitude.

While we leave the treatment of chaotic systems for the future, for now we study a highly idealized system of spherical symmetry, a spherically symmetric conglomeration of galaxies falling freely with spherical symmetry under gravity, all directed

---

[1] We remind the reader of the technical language that we use in Relativity: "Stationary" in General Relativity does not necessarily mean "not moving". It just means that there is no explicit time-dependence in the description of the motion. For example a disk that is perfectly axially symmetric and spinning at a constant rate about its symmetry axis, presents a totally time-independent picture to the observer. The distribution is moving but in terms of its density as a function of position, it is not varying in time. Such was the nature of our idealized model in the previous chapter. "Static" is the special case of "stationary" where the meaning is not only time-independent but also "not moving".

F. I. Cooperstock and S. Tieu, *Einstein's Relativity*,
DOI: 10.1007/978-3-642-30385-2_10, © Springer-Verlag Berlin Heidelberg 2012

towards the center-point of the distribution. As in our study of stationary systems, the system of individual concentrations of galactic matter is idealized as a continuum of pressure-free matter, i.e. "dust".

10.2 Perceptions of Velocity: Free-Fall in Vacuum

A key aspect of Relativity centers around the observer's perceptions of distance and time. In Special Relativity we focused on how observers in relative motion perceive length intervals and time intervals differently. In General Relativity, the differences were manifested as a result of observers being in regions of different gravitational field intensities, or in the new vocabulary, in different regions of spacetime curvature. With velocity being the distance covered divided by the time interval for it to be covered, clearly these differences of perception carry over into velocity as well. Before we get to the point of examining astronomers' perceptions of the velocities of galaxies in our idealized Coma Cluster, we turn to a very familiar subject to people studying General Relativity, the tracking of a small mass falling radially in vacuum towards the center of a spherically-symmetric mass. The detailed treatment of this subject is admirably covered in [3]. Here, we will aim to capture the essence of the phenomenon with the minimum of mathematics. This will provide us with the natural lead-in to the galaxy cluster model at hand.

Recall that when we discussed the nature of proper physical measurement of space and time, the metric tensor came into play. For example, the infinitesimal proper radial distance for the Schwarzschild spacetime with metric (5.4) is $dr/\sqrt{1 - 2m/r}$ at radial coordinate position r and the proper time interval is $dt\sqrt{1 - 2m/r}$. As a result, for the infinitesimal coordinate increments dr, dt, this observer judges radial velocity v_r as the ratio of these quantities, which, after simplification, is

$$v_r = \frac{dr}{(1 - 2m/r)dt}. \tag{10.1}$$

However an observer very distant from the mass, views spacetime geometry in line with the metric (5.2). For him, the increment of radial distance is dr and the time increment is dt. r is radial distance for him and t is time for him. As a consequence, the radial velocity v_r for him is

$$v_r = dr/dt. \tag{10.2}$$

When we solve the equations for free motion of a particle released from rest at a very large r value, the solution is found to be

$$\frac{dr}{dt} = -\left(1 - \frac{2m}{r}\right)\sqrt{\frac{2m}{r}}. \tag{10.3}$$

In line with (10.2), this is the velocity that the distant observer assigns to particles at any given r value. We see that for r being very large, the factor $\sqrt{\frac{2m}{r}}$ in (10.3) makes the perceived velocity value approach zero as a consistency check. Furthermore, as r approaches $2m$, the first factor in (10.3) makes the perceived velocity approach zero. Thus from the viewpoint of the distant observer, the falling particle effectively stops and never actually crosses the "event horizon" at $r = 2m$.

The picture is very different from the point of view of an observer who is stationed adjacent to the falling particle. For him, the perceived velocity is given by (10.1) and when we substitute the requisite dr/dt from (10.3) into (10.1), the result is

$$v_r = -\sqrt{2m/r}. \tag{10.4}$$

This is the complete opposite of the distant observer's view: as r approaches $2m$, from (10.4), we see that the adjacent observer views the falling particle's velocity approach ever closer to -1. Since we have assigned the value 1 to the speed of light for this coordinate system, we see that relative to the adjacent observer, the particle is approaching the speed of light with inward direction of motion (the minus sign).[2] The particle does not actually attain the speed of light since, as Synge had shown, the falling particle's crossing of the $r = 2m$ surface occurs within the light cone.

All of this is well-known to most relativists. We include it here both for its own sake as it is likely that the average reader is seeing it explained in this way for the first time and as guidance for the next step where we apply this formalism to the system of a collapsing spherical dust cloud.

10.3 Spherical Dust Collapse

Over the years, there has been much interest and a great deal of effort has been expended in studying the spherically-symmetric collapse of a cloud of dust. There were good reasons for doing so. First, spherical symmetry greatly simplifies analysis, particularly in General Relativity where this symmetry guarantees the absence of the complicating factor of gravity-wave emission. Second, the added factor of choosing dust, the zero-pressure fluid, adds a further layer of simplicity. Third, this model affords the luxury of yielding exact solutions, an advantage not to be under-estimated. This study was initiated in the 1930s by J. R. Oppenheimer and his collaborators, leading to the concept of the black hole. As mentioned earlier, we attacked the problem from a more realistic point of view, noting that with the densities mounting precipitously during gravitational collapse, the assumption of the continued mainte-

---

[2] For observers positioned closer and closer to $r = 2m$, it becomes more and more difficult for them to remain at rest relative to the central body. They require greater and greater rocket thrust directed towards the central body to do so. At $r = 2m$, an infinite amount of thrust would be required, underlining the impossibility of doing so. This explains why, on the one hand, the local velocity of the falling particle is seen to be approaching the speed of light as r approaches $2m$ yet there is no contradiction with the demand that non-zero rest-mass particles can never attain the speed of light.

nance of zero pressure is physically unrealistic [10]. Accordingly we considered the addition of pressure P to the model of gravitational collapse with material equation of state $P = k\rho$, where ρ is the density and k is a constant. We found the criteria for the eventual formation of black holes and naked singularities, the latter being, as the name implies, singularities that are not hidden from view by a horizon.

All of these studies had the goal of dealing with very strong gravitational fields. Up until the time that we did our more recent work on galactic motions, researchers would not have felt the motivation for studying the gravitational fields of spherically-infalling matter during the phase when the gravitational field was weak. To that point, the common wisdom was that Newtonian gravity is well-suited and surely adequate for that task. Einstein's gravity, it was believed, could only add insignificantly minor corrections. However, having seen with the stationary galaxy model that the common wisdom is limited in scope, we were motivated to examine the weak-field dust collapse problem more closely, looking ahead to the aim of modeling an idealized cluster of galaxies.

The complicated set of exact non-linear differential equations for a collapsing spherically-symmetric ball of dust as well as the exact solution to these equations is given in [3] (see also [1]). Given the complexity of the equations, the simplicity of the solution is quite remarkable, a tribute to the advantage of employing spherical symmetry. However, the extraction of the velocity dr/dt of the dust as viewed by distant observers, is a more difficult technical task. We have found this velocity, whose exact form is [24]

$$\frac{dr}{dt} = -\frac{(\alpha + \beta)(1 - \beta^2)}{8\pi r^2 \rho^2}\left[\frac{\alpha}{2M} + \beta\left(\frac{M''}{2(M')^2} - \frac{1}{4M}\right)\right]^{-1}\frac{\partial\rho}{\partial t} \tag{10.5}$$

where

$$\alpha = \frac{rM'}{3M}, \quad \beta = \sqrt{\frac{2M}{r}}. \tag{10.6}$$

In this expression, which is far richer than that deduced on the basis of Newtonian gravity,[3] the symbol M' indicates the rate at which the mass up to radius R changes with radial distance and M'' is the rate at which M' changes with radial distance. The density is ρ. With the new dependency factors now involved, there is scope for greater possibilities than had existed under the Newtonian gravity assumption.

We were able to show that within this model, the velocities of the galaxies as observed in the Coma Cluster were consistent with a total mass measure that does not include the vast extra mass requirement as computed on the basis of Newtonian gravitational theory. Thus, we see another layer to the power of Einstein's gravity to account for higher-than-expected velocities with lower-than-expected amounts of mass.

---

[3] The Newtonian-based expression depends only upon the value of the amount of mass within the sphere beneath the observation radius and the observation radius itself.

10.4 Summing Up the Dark Matter Picture

The idea that the material content of the universe is dominated by a mysterious kind of matter, now labeled "dark matter", that shows its presence through gravitational effects arose from Zwicky's observations of the galaxy motions within the Coma Cluster in the 1930s and Rubin's observations of the stars within galaxies in the 1970s. Dark matter is generally regarded as constituting approximately 95 percent of the matter in the universe and hence the specific understanding of its nature is justly regarded as one of the most fundamental issues in present-day physics. In turn the questioning of its very existence by a small group of researchers, the "questioners", including ourselves, ties in to the urgency of seeking out the essential truth of the matter, if the pun can be excused. While the other questioners have followed non-conservative tracks in their researches, it must be said that we have been uniquely conservative in our dynamical analyses, applying the most trustworthy gravitational theory known, namely Einstein's General Relativity. In doing so, we should be seen to be even more conservative than the researchers of the status-quo majority who profess unwavering faith in the reality of dark matter. After all, the latter base their findings on Newtonian gravitational theory with its known limitations.

So it all boils down to the issue of limitations. It is repeatedly claimed that for systems with weak gravity and non-relativistic motions, Newtonian gravity is perfectly adequate, that the theory's limitations are irrelevant in the domains of galaxy and galaxy-cluster dynamics. Yet in our simple model of a spiral galaxy, a model that incorporates weak gravity and non-relativistic motion, we have seen that the essential field equations are non-linear. With these field equations, we have demonstrated the higher-than-expected velocities can be accounted for with modest amounts of matter, obviating the assumed need for vast quantities of dark matter to encircle the galaxies in massive halos. An essential point is this: *researchers agree that the mathematics of non-linear differential equations is very different from the linear kind and in all the criticism that we have had directed against our work, there has never been any suggestion that we have erred in the correctness of our equations.* As well, in our modeling of a cluster of galaxies, there has never been any questioning of our application of the well-known procedure of velocity deduction by external observers.

Be that as it may, let us further review the current ideas pro and con framing the dark-matter hypothesis. Up to 95 percent of a galaxy's mass is conjectured to be in the form of dark matter by the dark-matter proponents. This measure is based upon an application of the Newtonian virial theorem to the diffuse interstellar gas at the galaxy's edge. However we do not yet have the General Relativity equivalent of the Newtonian virial theorem and we know that General Relativity can alter the picture remarkably. Furthermore, according to the dark matter theory of galaxy formation, there are 10 to 100 times as many small galaxies predicted to exist than are observed.

Gravitational lensing observations are frequently proffered as supporting the dark matter hypothesis. In this phenomenon, light from distant background objects such as galaxies is bent by intermediate masses on its way to observers on Earth. The gravity from the intermediate objects is the agent for the bending and as a result,

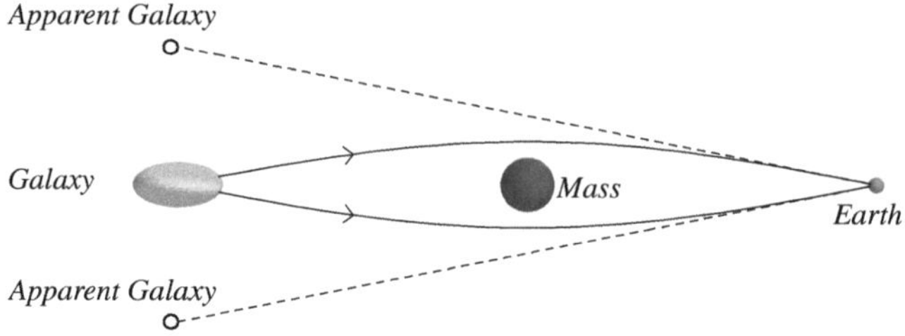

Fig. 10.1 Light from a star in a distant galaxy is viewed multiple times at different positions in the sky

the same distant galaxy can be viewed multiple times and seeming to be located at a variety of positions in the sky. In what is termed "strong lensing" stemming from large intermediary masses, the background galaxies appear to be distorted into arc shapes by the gravitational lens effect. In "weak lensing", a large number of very small distortions due to small intermediary masses, are observed and statistically analyzed. It has been claimed that General Relativity has been applied for the strong lensing and the results are consistent with the presence of dark matter. However the effects of rotation of the intermediary masses has not been taken into account and we have witnessed the importance of rotation in General Relativistic analyses.

The Bullet Cluster has been invoked as support for the dark-matter hypothesis. This cluster is regarded as the end-state of a collision between two previously separated clusters and is claimed to harbour large quantities of dark matter outside the central visible region. The idea is that while normal electromagnetically interacting matter was braked in the course of the collision and remained within the bounds of the collision site, the dark matter, immune from electromagnetic interaction, continued to follow its initial course and be located beyond the collision site, essentially oblivious to the effects of the collision. This is often seen as evidence for the existence of a new kind of matter. However it is amusing to note that subsequently, a new post-collision source, Abell520 [25] has been observed and analyzed. This source is seen to exhibit the exact opposite behaviour with a large dark massive region in the *center* of the collision region!

Various researchers claim that the strongest evidence in favour of the dark matter (and dark energy) hypothesis derives from the observed density fluctuations in the CMB (cosmic microwave background) observations. The claim is that these observations point to a $k = 0$ spatially flat (or at least very-nearly flat) universe. Yet we recall that current observations point to a far-from-adequate amount of normal baryonic matter in the universe to provide the necessary density for this flatness. While the estimated numbers vary with advances in research and observations, the rough values are 73 percent dark energy, 23 percent dark matter and a mere 4 percent normal matter. Later we will discuss the dark energy hypothesis but here we are concentrating on dark matter. Researchers are seldom at a loss to conjure hypotheses and for dark matter, they have conjectured three varieties: "hot dark matter", "warm dark matter" and "cold dark matter".

Hot dark matter, ultra-relativistic particles, is the least-favoured of the three hypothesized forms for dark matter because it is inconsistent with the currently favoured galaxy formation theory after the big bang. For this theory, slower particles are required to enable a clumping to occur to provide seeds for galaxy formation. It is ironical in that while the hot dark matter hypothesis is least favoured, such dark matter is the only form of the three types with *known* presence in the universe. This is in the form of neutrinos, the nearly massless highly relativistic leptons that are produced when neutrons decay into protons and electrons. These particles (as well as their more massive cousins, the mu- and tau-neutrinos) are not present in sufficient numbers to fill the large conjectured mass demand for dark matter.

More favoured and supported by a very small cadre of researchers is the warm dark matter conjecture. These warm dark matter particles are believed to be of very high velocities but not beyond the capacity to provide the necessary clumping for galaxy seed requirements. Since there are no such particles in the Standard Model of particle physics, the warm dark matter particles are conjectured to be found in the framework of "supersymmetry", the theory that for every lepton in nature, there exists a massive supersymmetric baryonic partner and for every baryon in nature, there exists a massive supersymmetric leptonic partner. This enables nature to recover its symmetry. In various universities in Europe, the theory of supersymmetry is considered so natural that it is taught as if it were established reality. In the case of warm dark matter, the hypothesized particles are the supersymmetric partners of the photon and the graviton, referred to as the "photino" and the "gravitino" respectively.

The most favoured form of dark matter is "cold dark matter", CDM. This is conjectured to consist of massive particles that are non-relativistic at the early stages of the universe evolution, thus enabling the clumping to form seeds for galactic cores. The COBE and WMAP satellites have revealed very small deviations from perfect isotropy in the very early universe, the evidence that is presented for the existence of such galactic formation seeds. Built into the assumption of the existence of cold dark matter particles is that prior to the appearance of the small deviations, there were no small deviations. We have no knowledge of the nature of the period before the maps revealed by COBE and WMAP so this is a form of compounded speculation. Particularly troubling for the cold dark matter conjecture is that the required particles, the so-called WIMPS, or weakly-interacting massive particles, have suffered two recent setbacks in their quest for revelation. First, to compose the required density of dark matter, there should be trillions of WIMPS passing through the Earth each second as it follows its path around the Sun. In ordinary circumstances, the WIMP collisions with normal matter would go undetected as the very few collision products would be noticed. However an experiment, the Xenon100 experiment, has been in operation in a deep mine in Gran Sasso, Italy. A cylinder of xenon has been suspended in vacuum in the depths of the Gran Sasso mine. Collisions of WIMPS with xenon nuclei would produce detectable emergent particles. This experiment has recently reached a very high level of sensitivity and to date *no WIMP-xenon collisions have been detected*; this in spite of the vast multitude of collisions that should have occurred had there been WIMPS present in the first place.

Even more recently, a second experimental result has been announced at CERN. This involves B-mesons in the LHC, the large-hadron collider. The result of this experiment has pointed to a failure of the theory of supersymmetry, a major blow to dark matter theorists. In our view, it is now time for researchers to turn back to well-established physics, in particular to Einstein's General Relativity and to probe further into the seemingly mystifying phenomena that might be explained more naturally using the favoured theory of gravity.

Chapter 11
Dark Energy

In very recent times, a new major issue has arisen: Is the true story of the very distant galactic motions, that is in terms of the universe as a whole, the one that had been assumed? Is it simply a matter of gravity slowing down the expansion of the universe? It certainly seems reasonable. However in more recent years, data has been presented to suggest (and most recently claimed to affirm) that rather than slowing down, the expansion of the universe is actually speeding up! This data comes from the observations of certain supernovae, massive exploding stars in distant galaxies. The observations of these distant supernovae are dimmer than they would have to be on the basis of a universe that is decelerating in the course of its expansion. Instead, the currently favoured view is that the universe is now undergoing an expansion *with acceleration*. This runs counter to our fundamental experience with gravitation, that gravity attracts rather than repels, the latter going hand-in-hand with acceleration accompanying the expansion.

How can physics account for this? Not easily, to be sure, and there remains considerable skepticism regarding the claim of accelerated expansion, this in spite of the recent awarding of Nobel Prizes for the observational evidence leading to the accelerated expansion conclusion. To introduce the "cure", let us return to Einstein's early view of the universe, the static universe. We know that gravity is attractive and free bodies do not want to stay put relative to each other. To remain at rest, they require outside assistance. Einstein provided this with his so-called "cosmological constant" Λ, a term that he added to his field equations of General Relativity, in the form

$$R_{ik} - \frac{1}{2} g_{ik} R + \Lambda g_{ik} = \frac{8\pi G}{c^4} T_{ik}. \tag{11.1}$$

This Λ acts as a repulsion between bodies to keep them at rest relative to each other. Einstein saw this Λ term as part of the spacetime geometry, belonging on the left geometry side of his famous equations. However, as we know from our algebra, a term on one side of the equation can be readily moved to the other side as long as one introduces a minus sign. It remains the same equation in essence but here the interpretation is important. In our earlier work (see for example [1]), we argued that

F. I. Cooperstock and S. Tieu, *Einstein's Relativity*,
DOI: 10.1007/978-3-642-30385-2_11, © Springer-Verlag Berlin Heidelberg 2012

Einstein's Λ term really belongs on the right-hand "matter" side of his equations, and then the Λ term is most naturally interpreted as a kind of very weird matter. It is matter that provides tension, and for Einstein, it had provided the right amount of tension to keep his universe static, opposing the gravitational attraction. After learning about the galactic red-shifts, Einstein abandoned his Λ. It had been widely reported that he regarded the introduction of Λ as the worst mistake of his life.

The great physicist, R. P. Feynman once remarked to us that physicists are not sufficiently imaginative. (Actually he used a more colourful expression but that was the idea.) Once the Λ genie was out of the bottle, it could not be put back in as Einstein had wished, at least not for long. Seeing the assumed need for accelerated expansion, why not bring back our old friend Λ? After all, it did the trick for Einstein when he needed it. Thus it is now widely believed that the universe is pervaded by a tenuous material called "dark energy" that provides accelerating thrust to the matter of the universe as required by the demands of the current data from supernovae. And Λ is the prime candidate to do the job of dark energy.

But the situation is more complicated than this. The current upper-limit to the magnitude of Λ is a very tiny number. After all, we recall how well Einstein's theory *without* any Λ at all had worked to give us the required 43 s of arc extra precession for the orbital motion of the planet Mercury. Too large a cosmological constant would have destroyed this wonderfully reliable result. But quantum field theory, that very successful theory of the micro-world, demands that the vacuum is not really without energy, but rather is imbued with an incredibly enormous amount of energy. This level of energy translates into a value that is actually 10^{120} times as large as the upper limit demanded by the observed value! This is not a misprint. It is not meant to be a hundred times or a thousand times or even a million times larger. It is worse- it is 1 followed by 120 zeros times larger! It makes a googol look like a very tiny number. Some have rationalized this by proposing that Einstein's theory provides a "bare" λ (note uncapitalized) which itself is enormous, almost exactly the same magnitude as that demanded by quantum field theory and that these two enormous numbers cancel each other so incredibly accurately as to leave behind the mere wisp of the observed net Λ of today! Most researchers, ourselves included, regard this idea as rather unbelievable.

Over the years, a number of alternative ideas have arisen. A variety of researchers have proposed that the cosmological term is not constant at all, but rather it evolves in tune with the evolution of the universe in such a manner as to be large when needed to be large and to be consistent with the present value today. (See [32] for a review of these many forms as well as some new forms). The problem with this approach is one of rationalizing the form of this variation on a fundamental level.

An intriguing alternative approach was provided in [33]. These authors argued that metallic vapours are ejected in the course of the kind of supernovae explosions that have indicated accelerated expansion and that these vapours are pushed out of the galaxies in which the supernovae reside, by shock waves. Such debris would have the tendency to diminish the intensity of the radiation that would ordinarily come directly to us, thus explaining the dimmer supernovae without any need for an accelerated expansion explanation. However others would argue that the CMB observations provide another avenue of support for the necessity of a cosmological term. Still others would argue that making deductions on the basis of these very-early-universe observations is akin to the reading of tea leaves. As you can now appreciate, research in cosmology is unlike the precision research related to laboratory experimental verification that one generally ascribes to scientific research.

So we see how various researchers have dealt with the observations. One, E. W. ("Rocky") Kolb has colourfully described dark energy as the ether of the 21st century. How this will all play out in time is of course a great unknown. But it is the unknown that keeps us engaged in the fascinating pursuit of understanding our miraculous universe.

Chapter 12
Time Machines, the Multiverse and Other Fantasies

12.1 Closed Time-Like Curves

It is our everyday life experience that we re-visit previous places. We take it for granted that we always do so at later times. Even with the wonders of Special Relativity, when we considered the experiences of Alicia and Beatrice in Chap. 3, we saw that Beatrice re-visited her old home at a later time for her and the re-visit was at a much later time for Alicia, so much later in fact, that she was no longer a member of the living. We can accept such phenomena, however incredible they may seem, because they do not violate the demands of causality, that if A causes B, A will always be seen to precede B in the ordering of events in spacetime. While the perceived time differences for Alicia and Beatrice are very different as a consequence of Special Relativity being in accord with the workings of nature, Special Relativity maintains the necessary requirement of causality preservation. However, human imagination does not always display such constraints.

In works of science-fiction and with increasing frequency in the movies, we witness somewhat outrageous plots in which characters routinely take trips into their past and back to the future again, with often predictable consequences. Some plots skirt around the causality issue by having the time-traveler merely observe certain goings-on without being allowed to interfere with them. Significantly, even some prominent physicists have been swept along in the excitement of the idea. Some writers realize that their targeted audience might ask some embarrassing questions. For example, if the time-traveler, in the course of a stint during which he is re-visiting his past, were able to kill his grandmother in a fit of rage when she was just a very young child, then how could the time-traveler have ever been born to be able to be there to make the voyage in the first place? Good question.

The renowned mathematician, Gödel, had taken an interest in General Relativity and found a solution of the Einstein field equations of General Relativity for a rotating universe [34]. This solution, he excitedly proclaimed, exhibited what are called "closed time-like curves". Real observers follow time-like curves in spacetime, those physically permitted curves that lie within the observer's future light-cone, and if

F. I. Cooperstock and S. Tieu, *Einstein's Relativity*,
DOI: 10.1007/978-3-642-30385-2_12, © Springer-Verlag Berlin Heidelberg 2012

these curves are closed as a result of the light-cone tipping over sufficiently, it means that the observer, when his spacetime path touches the closure point, is re-visiting his past. Thus, Gödel reasoned, a person in this universe could be a time-traveler.

Gödel was so excited about this result that he gave a seminar at the renowned Institute for Advanced Studies in Princeton which was attended by such luminaries as Einstein, Oppenheimer and Chandrasekhar [35]. We can only wonder whether Gödel was seriously challenged as to the physical significance of his work, but at least Chandrasekhar took it sufficiently seriously to write a paper that made note of the fact that Gödel's closed time-like curves were not geodesics [36]. Now we recall that only freely gravitating particles follow geodesics, so Gödel's time-travelers would have required some assistance in the form of rocket propulsion or the like to get back into their past. While this might add some extra degree of complication, nevertheless the basic assertion that this time-travel experience could be an element of reality was maintained as a possibility to be reckoned with. Gödel's work did not garner much attention for many years until the appearance of a highly technical book that featured this work [37] (Fig. 12.1).

More recently, we set ourselves the task of considering both Gödel's solution and the entire issue of closed time-like curves. We came to a very simple conclusion: The realization of a closed time-like curve is a matter of a mathematical choice of the identification of spacetime points and has nothing to do with physics. Let us construct a very simple example. Suppose a person is walking around a big circle. At the start of the person's journey, a point marked x on the circle, it is 1 o'clock. One quarter of the way around the circle, his clock reads 1:15, half way around it is 1:30, three-quarters of the way around it is 1:45. When he finally reaches the starting point x, it is 2 o'clock. Now suppose we were to *identify* the start spacetime point $(x, 1\ o'clock)$ with the finish point $(x, 2\ o'clock)$, in other words to say that 2 o'clock equals 1 o'clock. We could then proclaim that the traveler has gone back in time! He has not only revisited his previous spatial position, a routine of our experience, but he has also revisited the same time at which he had been at the position earlier. Now you might object to such a cavalier statement of identification but there is nothing to prevent one in terms of mathematics from making such an identification.

Now when the traveler went from the 0° position in the circle at 1 o'clock to 90° at 1:15 to 180° at 1:30 to 270° at 1:45 to end up at the starting *spatial* point x, we have no problem in identifying the final angular value of 360° with the start value 0°. This is our experience in nature, the revisiting of spatial points. And it is precisely this aspect that enters into the Gödel example.

It arises in the following manner: in Gödel's solution there is an angular coordinate ϕ which has the standard characteristic of identification, wherein the advance in angular position from 0 to 360° brings one back to 0, the original angular position. This does not raise any concern for us in those regions where ϕ is a space-like coordinate. However, in the Gödel solution there is a region where ϕ becomes a time-like coordinate. That is where the problem lies. Gödel, and researchers since then, have continued to regard ϕ in this time-like region in the same manner as they had when ϕ was space-like and have continued to regard 0 and 2π, i.e. 360°, as relating to the same spacetime point. In doing so, lo and behold, a closed time-like

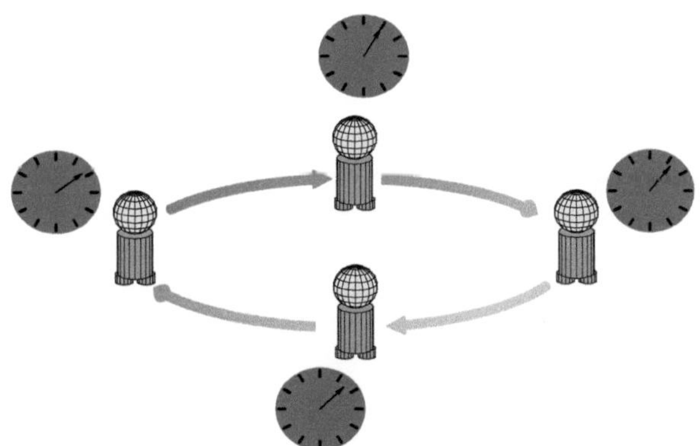

Fig. 12.1 A person walks around a *circle* carrying a clock. He begins at 1 o'clock at the *top* of the figure and completes one cycle at 2 o'clock. By identifying 2 o'clock with 1 o'clock, a closed time-like curve is realized

curve is born! It is important to emphasize that this is a *choice* that has been made. It is *not* a physical necessity.

Einstein's field equations lead to certain solutions. Hand in hand with a given solution, we assign a meaning to the coordinates of that solution, for example that r runs from 0 to ∞, that ϕ runs from 0 to 2π *with 0 and 2π being identified*, and z and t run from $-\infty$ to $+\infty$. This tells us that the coordinates have the character of cylindrical polar coordinates. The process in its entirety, the given solution to the field equations plus the characterization of the coordinates, gives us a particular spacetime. If, as part of the specification, we continue in this same vein even when ϕ becomes a time-like coordinate, we have Gödel's spacetime. However, when we choose instead *not* to identify the points in ϕ when ϕ becomes time-like, we have a different, and we would argue a more reasonably physical spacetime. It is as simple as that. Surprisingly, much has been made of the supposed inclusion of closed time-like curves in General Relativity and it is likely that this has served as a spur towards the infiltration of the bizarre time-machine plots in more recent fiction offerings. These aspects and various examples are developed in some detail in [38].

12.2 Topological Twists

As in various areas of human endeavours, physics has had its share of fads. In recent years, topological structures have come to the fore as supposed elements of physical reality. A favourite of this genre is the wormhole. The wormhole, as the figure shows, presents a means of getting from A to B by the alternate path of entering the wormhole and emerging from the other end. Those with a science-fiction bent have seized upon the wormhole as a means of drastically cutting the time required for space-travel between distant galaxies: find the wormhole in the spacetime that offers a greatly shortened travel-time requirement and space-travel becomes a feasible option for human beings.

Fig. 12.2 The normal path from A to B is along the straight portion from A, then around the flap at C, continuing to B. However, the shorter path from A to B is through the wormhole

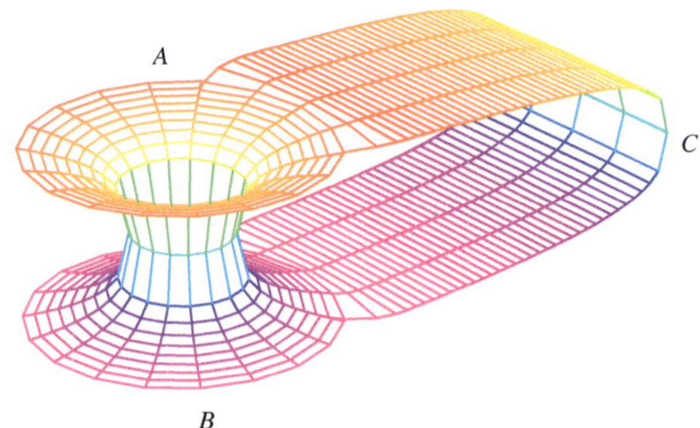

To get a rough picture as to how this works, consider a normal path in spacetime between points A and B. This is the course that follows the undistorted path from A around the flap at C and back to B. However, should A and B be connectable through a wormhole as in Fig. 12.2, one could connect A and B via the shortened path through the wormhole.

It becomes even more engrossing when the set-up is envisaged wherein the journey through the wormhole leads to travel to the observer's past, a newly-minted time-machine. Those of us who prefer to have science clearly separated from science-fiction are left perplexed as we witness actual scientists debating whether or not nature will intercede to pinch off the travel of any mischievous wormhole-sniffer who tries to enter the alluring wormhole time-machine.

We are left wondering: where do these scientists glean even the slightest hint that such topological structures are actually a part of physical spacetime? Without such, we would suggest that there is no motivation to reach in such directions. It is natural to ask for an explanation as to why a considerable body of discourse in science has verged towards science-fiction. Perhaps the most cogent explanation derives from the increasing difficulty to achieve experimental results. The experiment has always been the ultimate arbiter of truth. As theory demands greater and greater energies for experimental confirmation and as these energy scales become increasingly difficult (and expensive) to achieve, the disconnect between experiment and theory grows. In our view, it has grown to the point of irresponsible speculative pronouncements often motivated more by human than scientific needs.

12.3 The Multiverse

The picture of the universe that we have developed to this point precedes the radical theorizing developments that have accrued over the past two decades. The earlier picture was based upon our most trustworthy theory of gravity: General Relativity, its homogeneous isotropic solutions incorporating isotropic expansion, in line with our observations of the distant galaxies. As well, the solutions provide that any observer would witness this same isotropic expansion, a reasonable assumption. After all, otherwise one would ask: "Why should we be so lucky as to see an isotropic expansion?" We are just ordinary observers circling an ordinary star in an ordinary galaxy populated by billions of similar galaxies. While the solutions provide for the possibility of three kinds of basic form, an observable criterion for the right selection exists with the determination of average matter density. This is science at its best, elegant substantiated theory with exact mathematical solutions leading to observable verification within our reach. Unfortunately, the more recent cosmological developments have not followed the scientific path, as we will discuss below.

The first sign of trouble is in vocabulary. The word "universe" refers to the totality of all that exists. Proponents of the new cosmologies hypothesize that just as there are many galaxies, there are many universes, collectively labeled the "multiverse". However, since the universe is already the totality of all that exists, introducing the new word "multiverse" is most charitably seen as an act of injecting redundancy. The proponents of the multiverse idea might object in that their definition of our universe is all that is within our cosmic visual horizon, the furthest distance to which we could possibly probe, based on the distance that light could reach us from the time of the big bang. Their other universes comprising the multiverse are those which are never within our possibility of observation. However, let us use their new vocabulary and briefly describe a few of the various options being proposed.

What one might regard as the conservative among the new cosmologists accept the multiverse with many, if not infinitely many universes much like our own, obeying the same laws of physics that we observe to hold sway. On the other hand, one might view as the more venturesome cosmologists, those who propose the possible existence of many or an infinite number of other universes obeying different physical laws and even different numbers of dimensions, even many with copies of ourselves (the "parallel universes" idea). However, since the external universes of both the conservatives and the more adventurous are all forever beyond our reach, both groups of cosmologists have a key element in common: Neither one is engaged in the activity that we would normally call "science" but rather in the sphere of science-fiction. This is because their theorizings are a priori not verifiable.

Some have been very impressed by the values of the fundamental constants of nature vis-a-vis their unique suitability for the existence of life. They argue that the existence of this special set points to the viability of a multiverse with individual copies having a wide spectrum of values for the fundamental constants. With a wide choice available, to their way of thinking, the existence of one with the crucial values becomes plausible. Implicit in this line of reasoning is that the multiverse repeats structures with physical laws of our own, differing only in the numerical values of their physical constants. But how are we to know? A means of testability is called for.

Some cosmologists have even skirted with the idea that they could bring their speculative theorizing closer to real science by seeking evidence in the cosmic background radiation for a prior collision of an alien universe with our own. Seeking evidence is welcome news indeed. However, with the essential uncertainty (or perhaps "unknowability"!) of what preceded the big bang, one could hardly rule with any degree of confidence as to what were the sources of any anomalies found in the background radiation. Could they derive from our universe or from an alien universe? Who's to say?

Chapter 13
Concluding Commentary

Very recently, a new observation has been announced: A burst of neutrinos has been generated at the CERN facility in Geneva and directed through a slice of the Earth's crust and towards the Gran Sasso mine in Italy. Remarkably, the experimenters have measured the speed of the burst as travelling slightly faster than the speed c of light in vacuum. Assuming this measurement was accurate, and that it harboured no experimental flaws,[1] this is indeed a very stunning result for different reasons.

First, the entire structure of Relativity rests upon c as the universal speed limit in nature and that only zero-rest-mass particles such as photons can travel at speed c, and this, only in vacuum. Moreover the theoretical structure of Lorentz-invariance is built into the entire theory and the theory works brilliantly. Second, in recent times, it has come to be accepted by particle physicists that neutrinos, even the mundane variety of electron-neutrinos, actually do have some mass, albeit very little and the muon-neutrinos and tau-neutrinos have even more so. We recall that according to Special Relativity, it would require an infinite amount of energy to make a non-zero-mass body reach the speed of light c, let alone exceed it.

What are we to make of this truly remarkable claim from Europe? To begin, as the claimers have said responsibly, attempts should be made to replicate their claim at different sites in different experimental configurations. Let us assume it is replicated. What does it say to us? In our view, it points to a general problem that has festered for a very long time, that of our limited understanding of the relationship between the micro-world and the macro-world, between the world of quantum mechanics, quantum field theory and macroscopic Relativity. With electromagnetism, at least we have the well-defined and well-established classical theory of Maxwell. We deal with this in our daily lives in so many ways. As well, in those areas where the classical theory is left wanting, the quantum theory of electromagnetism comes to the rescue. This theory offers extremely precise verification even though it leaves the issue of vacuum energy unresolved and some irritating infinities carefully tucked

---

[1] The latest news suggests that a loose cable was responsible for displaying a false reading of the speed.

F. I. Cooperstock and S. Tieu, *Einstein's Relativity*,
DOI: 10.1007/978-3-642-30385-2_13, © Springer-Verlag Berlin Heidelberg 2012

away. The picture of the quanta of the electromagnetic field, the photons, fit well into our appreciation of the nature of electromagnetism at the quantum level. The photoelectric effect, the spectrum for black-body radiation and the Lamb-shift of the spectral lines in hydrogen underline this excellent fit.

By contrast, we do not have such a familiarity with neutrinos. They were brought into the physical fold as the required quantized particles to rescue the conservation of energy-momentum and angular momentum in phenomena such as the decay of a neutron. At first they were seen to be the massless spin-1/2 analogues of the spin-1 photons but more recently, were said to have mass. The theory evolved to the form wherein the neutrinos of the different kinds, the e-, the μ- and the τ-neutrinos transform into each other under the process referred to as "neutrino oscillations", in the course of their propagation. Now we have the latest claim of their speed being greater than c. All of these results have come in rapid succession by the standards of fundamental advances in physics. On top of all of these results, we have to keep in mind that unlike photons with their classical electromagnetic wave connection, we have no classical neutrino phenomena to back up our confidence in their innate character.

While all of the foregoing is very interesting and the recent observations potentially of epic importance, it is time to step back and take a breath before we jump to conclusions. Our approach from the outset has been the conservative one, in line with Occam and his razor. We see the enormous challenge inherent in the incorporation of dark energy in line with the demands of the vacuum in quantum field theory. As well, we ask whether more mundane explanations for the apparent acceleration of the expansion of the universe have been adequately pursued. And of particular interest for us, we reflect upon the strong resistance mounted against our displays that General Relativity can account for the motions of the galaxies, both individually and in ensembles, with little or no dark matter. We marvel at how General Relativity almost magically predicts the optical edge of galaxies simply from the knowledge of the galactic rotation curves. This is an achievement unattainable within Newtonian gravity theory. And this achievement underlines the necessity of applying General Relativity to galactic dynamics.

History has shown us that the evolution of sober scientific thought often proceeds at a snail-pace while facile "fixes" are often embraced rapidly with insufficient critical analysis. Eventually the truth emerges.

Appendix A:
Proving That the Spacetime Interval
is an Invariant

Our proof is along the lines of that in [3] but it is somewhat simpler. Noting that when $ds = 0, ds^* = 0$ and that these elements are infinitesimals of the same (first) order, they must be proportional. On what can the proportionality factor depend? It cannot depend on where one is located in space because no point in space is privileged over any other point. Space is assumed to be homogeneous. It cannot depend upon the time because no point in time is privileged. Now we turn to the only element that relates Alicia to Beatrice; it is their relative velocity v. Now it cannot depend upon the direction of their relative velocity because no direction is favoured over any other direction; the spacetime is isotropic. All that is left is the magnitude of their relative velocity, $\mathrm{mod}(v)$. So we write

$$ds = K(\mathrm{mod}(v))\, ds^* \tag{A.1}$$

where K is the proportionality function. Suppose we were to consider the relationship from Beatrice's viewpoint. For her, Alicia has velocity $-v$ relative to Beatrice so the connection between the intervals is

$$ds^* = K(\mathrm{mod}(-v))\, ds. \tag{A.2}$$

But $\mathrm{mod}(v) = \mathrm{mod}(-v)$ so the K functions in (A.1) and (A.2) have the very same value. Thus a substitution of ds^* from (A.2) into (A.1) gives

$$ds = K(\mathrm{mod}(v))\, ds^* = K(\mathrm{mod}(v))K(\mathrm{mod}(-v))\, ds = K^2\, ds \tag{A.3}$$

and therefore $K^2 = 1$. When we square both sides of (A.1) and substitute $K^2 = 1$, we have

$$ds^2 = ds^{*2}. \tag{A.4}$$

Thus $ds = (+/-)ds^*$ and we choose the plus to preserve the direction of the flow of time. From the equality of the infinitesimals, it follows that the finite intervals are equal as well, $s = s^*$.

The important result that emerges is that the spacetime interval between any two given events is an invariant.

F. I. Cooperstock and S. Tieu, *Einstein's Relativity*,

DOI: 10.1007/978-3-642-30385-2, © Springer-Verlag Berlin Heidelberg 2012

Appendix B:
Deriving the Einstein Field Equations

In Newtonian gravitational theory, there is one simple equation for the gravitational field that connects the gravitational potential ϕ to the distribution of matter density ρ. This equation is of the type that gives solutions for density changes that imprint upon the entire universe instantaneously. This is inconsistent with the basic premise of Special Relativity, so if Special Relativity is correct, Newtonian gravity must be replaced by a relativistic theory of gravity.

For those readers who are interested in developing a detailed understanding of the mathematics underpinning General Relativity, they would profit from reading the next level of treatment in [1] with extra tensor calculus detail in [7]. They could then follow up with more sophisticated treatments in the classic book of Landau and Lifshitz [3] as well as the various other more advanced books on General Relativity. In the present book, we are aiming at a very basic level of development with just enough mathematics to go beyond those books which rely solely upon descriptions in words.

Without getting into more advanced mathematical detail, we present a narrative that leads naturally to Einstein's General Relativity. To do so, we first consider the very basic constructs of energy and momentum. What makes them basic is that they have the valuable property of being conserved when systems undergo changes. Much of physics through the centuries has been driven by the new constructs that had been invoked to maintain this primary attribute. For example, in electromagnetism, we express the conservation of the important physical attribute that we label "charge" in the form

$$\nabla . J + \frac{\partial \rho}{\partial t} = 0 \tag{B.1}$$

where ρ here is the charge density, the amount of charge per unit volume and J is the charge-current density vector, ρv with v being the velocity of the charge at the point under investigation.

To see why (B.1) expresses the conservation of charge, we require the fundamental theorem of Gauss that the integral of the divergence of a vector over a

F. I. Cooperstock and S. Tieu, *Einstein's Relativity*,
DOI: 10.1007/978-3-642-30385-2, © Springer-Verlag Berlin Heidelberg 2012

given volume is equal to the flux of that vector over the surface that bounds the given volume:

$$\int (\nabla . J)\, dV = \int J.ndS \qquad\qquad (B.2)$$

where n is a vector of unit length perpendicular to the bounding surface and pointing outward at every point from that surface. When we integrate (B.1) over a given volume and apply (B.2), we have

$$\frac{d}{dt}\int \rho dV = -\int J.ndS. \qquad\qquad (B.3)$$

This equation states that the time-rate of change of the total charge within the volume V is given by the negative of the rate at which charge flows out through the bounding surface S. In other words, what is lost in the volume is necessarily compensated by what flows out of its bounding surface. Thus we see that (B.1) is an expression of conservation locally.

In Relativity, the natural arena for investigation is spacetime rather than space, and we generalize the three-dimensional current vector $J = (\rho v_x, \rho v_y, \rho v_z)$ to a new four-dimensional current vector J^i, $J^i = (\rho, \rho v_x, \rho v_y, \rho v_z)$, i.e. we add a fourth "time-component" ρ in going from the three-vector J to the four-vector J^i. In doing so, we are able to express the conservation Eq. (B.1) as

$$\frac{\partial J^i}{\partial x^i} = 0. \qquad\qquad (B.4)$$

This is a very useful and important result: The vanishing of the divergence of a four-vector (or tensor) expresses conservation.

In a similar manner, energy and momentum conservation in Special Relativity is expressed by the vanishing of the divergence of the "energy-momentum tensor" T^{ik}:

$$\frac{\partial}{\partial x^i} T^{ik} = 0. \qquad\qquad (B.5)$$

Here, instead of dealing with a vector, we have to use a tensor (of second rank-i.e. two indices i and k instead of the single index that represents a vector.) Actually a vector is also a tensor; it is a tensor of first rank. While the vector J^i has four components (i taking on the values $0, 1, 2, 3$), a second rank tensor has 16 components in general but because this tensor is symmetric, i.e. $T^{01} = T^{10}$, $T^{12} = T^{21}$..., etc., only 10 components of this energy-momentum tensor are distinct in general.

Now let us consider working in Special Relativity using non-Cartesian coordinates, for example polar coordinates, or even transforming to a reference frame that is under acceleration. It is frequently stated that the latter case is no longer one of Special Relativity, that one is dealing with General Relativity whenever acceleration enters the picture. This is quite untrue. One can apply

Special Relativity in an accelerated reference frame. In fact Bondi [2] and others have done so. General Relativity enters the discussion when *real* gravity is present, namely spacetime curvature generated by matter and/or fields, not the pseudo-gravity of accelerated reference frames.

So let us suppose that we are working within the domain of Special Relativity but not in an inertial Cartesian system of coordinates. The formalism of tensor calculus tells us how the usual partial derivative of a vector or tensor transforms (see, e.g. [1, 7]). The usual partial derivative of a vector J^i for example, changes to what is called the "covariant derivative", denoted by a semi-colon. Specifically,

$$J^i_{;k} = \frac{\partial J^i}{\partial x^k} + \Gamma^i_{lk} J^l \qquad (B.6)$$

where

$$\Gamma^i_{lk} = \frac{1}{2} g^{im} (g_{ml,k} + g_{mk,l} - g_{lk,m}) \qquad (B.7)$$

and we have used commas to indicate the usual partial derivatives. Thus we see that the metric tensor comes into play when we deviate from simple Cartesian coordinates.

We now have the means of expressing the conservation of charge and the conservation of energy-momentum in terms of arbitrary coordinate systems in Special Relativity. Specifically, charge conservation changes from (B.4) to

$$J^i_{;i} = 0 \qquad (B.8)$$

and the expression for energy-momentum conservation changes from (B.5) to

$$T^{ik}_{;k} = 0. \qquad (B.9)$$

Now locally, an accelerated reference frame is like a gravitational field (Equivalence Principle) so we are led to the expression for energy-momentum conservation in General Relativity: simply (B.9). There is really no other possibility!

We are almost there. Recall that the mass density ρ is the source of gravity in Newtonian physics and most often, when we wish to go to the limit of weak gravity and non-relativistic velocities, we should retrieve Newtonian gravity from the relativistic theory of gravity that we seek. Interestingly, the T^{00} component of the energy-momentum tensor T^{ik} is ρ in the reference frame in which the matter is at rest so we seek a relativistic gravity field equation of the form

$$G^{ik} = \text{constant}.T^{ik}. \qquad (B.10)$$

Thanks to our earlier discussion regarding the conservation laws, we know what to look for in this so-far undetermined tensor G^{ik}. We know that since T^{ik} satisfies (B.9), so too must the right hand side of (B.10), i.e.

$$G^{ik}_{;k} = 0. \qquad (B.11)$$

As well, G^{ik} must be constructed from g_{ik} and no higher than its second derivatives so that in the appropriate limit, the left hand side of (B.10) reduces to the "Laplacian" of the Newtonian potential ϕ, i.e. $\nabla^2\phi$ where $\nabla^2 = \partial^2/\partial x^2 + \partial^2/\partial y^2 + \partial^2/\partial z^2$ when we use Cartesian coordinates.

Einstein worked diligently in search of the tensor G^{ik} and in the early stages of his search, he believed that the correct form of G^{ik} is the Ricci tensor R^{ik}. This tensor is a very complicated and lengthy combination of Christoffel symbols and their derivatives:

$$R_{ik} = \Gamma^n_{ik,n} - \Gamma^l_{il,k} + \Gamma^l_{ik}\Gamma^n_{ln} - \Gamma^n_{il}\Gamma^l_{kn}. \tag{B.12}$$

Unfortunately, this tensor does not have a vanishing covariant divergence as it must to be satisfactory. Eventually Einstein discovered that the required tensor is

$$G^{ik} = R^{ik} - \frac{1}{2}g^{ik}R^j_j, \tag{B.13}$$

now appropriately named the "Einstein tensor". Thus we have developed the Einstein field equations of General Relativity,

$$R^{ik} - \frac{1}{2}g^{ik}R^j_j = \frac{8\pi G}{c^4}T^{ik} \tag{B.14}$$

where the constant $\frac{8\pi G}{c^4}$ has been chosen in order to reduce to the Newtonian field equation for gravity

$$\nabla^2\phi = 4\pi G\rho \tag{B.15}$$

in the limit when appropriate. Rather than one simple field Eq. (B.15), we have ten very complicated non-linear partial differential equations. There are in general, ten distinct equations, counting the permutations of $i, k = 0, 1, 2, 3$ and taking into account the symmetry, $T^{10} = T^{01}$, $T^{12} = T^{21}$, etc. Given the complexity, while the task of finding valuable solutions might appear hopeless, with sufficient symmetry, physicists have found some very valuable solutions. These are discussed in the text. More solutions can be found in the advanced texts.

Regarding motion in General Relativity, the Equivalence Principle again serves as a valuable guide. Recall from Newtonian physics, that when there are no forces or when the forces on a body balance out, the body moves with zero acceleration. This holds in Relativity as it does with Newton:

$$a^i = 0 \tag{B.16}$$

for zero force. Here, we have written the acceleration as having four components instead of three. This is reasonable as after all, the arena of investigation in Relativity is four-dimensional spacetime rather than three-dimensional space. You might wish to delve more deeply into the mathematics of Relativity in the suggested references where four-dimensional vectors and tensors are discussed.

Now suppose we were to examine the system from the viewpoint of an accelerated reference frame. Then the mathematical transformation of this vector a^i turns it into a different form, a^i_{intrin}, called the "intrinsic acceleration", in terms of the new coordinates, and it is equal to zero as before:

$$a^i_{\text{intrin}} = a^i + \Gamma^i_{kl}u^k u^l = 0. \tag{B.17}$$

Here, the Γ is the Christoffel symbol as discussed above and u^k is the four-vector velocity, appropriate to Relativity, just as a^i is the previous four-vector acceleration. From the Equivalence Principle, gravity is locally like being in an accelerated reference system. Therefore it is natural to take (B.17) to be the equation of motion of a free body in full-blown General Relativity, i.e. when the body is moving in curved spacetime, not only when it is moving under pseudo-gravity. But you might well object to the effect that when there is real gravity present, like the pull of the earth, it is no longer a force-free situation. This is because you are so conditioned to thinking of gravity as a force. General Relativity tells you that you must think differently about gravity. Gravity is now removed from the collection that we call "forces" and is identified with spacetime, that is, its attribute that we call "curvature". And when you ask: "How does a body move under this new expression for gravity when all the forces are absent or balanced out?", the answer is: "freely", according to (B.17), the "geodesic equation".

References

1. F. I. Cooperstock, 2009 *General Relativistic Dynamics* World Scientific Publishing, Singapore.
2. H. Bondi in *Lectures on General Relativity, Brandeis Summer Institute in Theoretical Physics* eds. S. Deser and K.W. Ford, Vol.1, Prentice-Hall, New Jersey, 1965.
3. L. D. Landau and E. M. Lifshitz, *The Classical Theory of Fields* Fourth revised English edition, Pergamon Press, Oxford, 1975.
4. R. P. Feynman, R. B. Leighton and M. Sands, *The Feynman Lectures on Physics*, Addison-Wesley, Reading, Mass., 1964.
5. W. Isaacson, *Einstein*, Simon and Schuster, New York, 2007.
6. J. L. Synge, *Relativity: The General Theory*, North-Holland, Amsterdam, 1966.
7. B. Spain,*Tensor Calculus*, Oliver and Boyd, Edinburgh, 1960.
8. R. Adler, M. Bazin and M. Schiffer, *Introduction to General Relativity*, McGraw-Hill, New York, 1965.
9. J. L. Synge, *Proc. Roy. Irish Acad.* **A53**, 83, 1950.
10. F. I. Cooperstock, S. Jhingan, P. S. Joshi and T. P. Singh, *Class. Quantum Grav.* **14**, 2195, 1997.
11. F. I. Cooperstock, *Ann. Phys. N.Y.* **282**, 115, 2000; F. I. Cooperstock and S. Tieu, *Found. Phys.* **33**, 1033, 2003.
12. M. Gurses and F. Gursey, *J. Math. Phys.* **16**, 2385, 1975.
13. W. B. Bonnor, *Commun. Math. Phys.* **51**, 191, 1976.
14. A. Papapetrou, *Ann. Phys. (Leipzig)* **20**, 399, 1957; **1**, 185, 1958.
15. F. I. Cooperstock, *Ann. Phys. N. Y.* **47**, 173, 1968.

16. H. Bondi, M. G. J. van der Burg and A. W. K. Metzner, *Proc. Roy. Soc.* **A269**, 21, 1962.
17. J. Madore, *Ann. Inst. Henri Poincare'* **12**, 365, 1970.
18. F. I. Cooperstock and D. W. Hobill, *Phys. Rev.* **D20**, 2995, 1979.
19. H. Bondi, *Proc. Roy. Soc. London* **A 427**, 249, 1990.
20. M. J. Dupre, gr-qc/0903.5225.
21. F. I. Cooperstock and M. J. Dupre, gr-qc/0904.0469; *Int. J. Mod. Phys.* **D19**, 2353, 2010.
22. R. C. Tolman, *Relativity, Thermodynamics and Cosmology*. Clarendon, Oxford, 1934.
23. F. I. Cooperstock and S. Tieu, astro-ph/0507619; astro-ph/0512048; *Mod. Phys. Lett. A.* **21**, 2133, 2006; *Int. J. Mod. Phys.* **A22**, 2293, 2007.
24. F. I. Cooperstock and S. Tieu, *Mod. Phys. Lett. A.* **23**, 1745, 2008.
25. A. Mahdavi et al, astro-ph/07063048.
26. A. S. Eddington, *Proc. Roy. Soc.* **A 102**, 268, 1922; *The Mathematical Theory of Relativity*, Cambridge Univ. Press, Cambridge, U.K., 1923.
27. W. J. van Stockum, *Proc. R. Soc. Edin.* **57**, 135, 1937.
28. W. B. Bonnor, *J. Phys. A: Math. Gen.* **10**, 1673, 1977.
29. B. Fuchs and S. Phleps, *New Astron.* **11**, 608, 2006.
30. J. D. Carrick and F. I. Cooperstock, astro-ph/1101.3224; *Astrophys. Space Sc.* **337**, 321, 2012.
31. S. M. Kent, *Astron. J.* **93**, 816, 1987.
32. J. M. Overduin and F. I. Cooperstock, *Phys. Rev. D* **58**, 043506, 1998.
33. S. M. Chitre and J. V. Narlikar, *Astrophys. Space Sc.* **44**, 101, 1976.
34. K. Gödel, *Mod. Phys.* **21**, 447, 1949.
35. I. Ozsvath and E. Schucking, *Am. J. Phys.* **71**, 801, 2003.
36. S. Chandrasekhar and J. P. Wright, *Proc. Natl. Acad. Sci. U.S.A.* **47**, 341, 1961.
37. S. W. Hawking and G. F. R. Ellis, *The Large Scale Structure of Spacetime*, Cambridge University Press, Cambridge, 1973.
38. F. I. Cooperstock and S. Tieu, gr-qc/0405.114; *Found. Phys.* **35**, 1497, 2005.

Index

F. I. Cooperstock and S. Tieu, *Einstein's Relativity*,
DOI: 10.1007/978-3-642-30385-2, © Springer-Verlag Berlin Heidelberg 2012

MIX
Papier aus verantwortungsvollen Quellen
Paper from responsible sources
FSC® C105338

If you have any concerns about our products,
you can contact us on
ProductSafety@springernature.com

In case Publisher is established outside the EU,
the EU authorized representative is:
Springer Nature Customer Service Center GmbH
Europaplatz 3, 69115 Heidelberg, Germany

Printed by Libri Plureos GmbH
in Hamburg, Germany